The Newsphere

This book is part of the Peter Lang Media and Communication list.
Every volume is peer reviewed and meets
the highest quality standards for content and production.

PETER LANG
New York • Washington, D.C./Baltimore • Bern
Frankfurt • Berlin • Brussels • Vienna • Oxford

Christine M. Tracy

The Newsphere

Understanding the News and Information Environment

PETER LANG

New York • Washington, D.C./Baltimore • Bern
Frankfurt • Berlin • Brussels • Vienna • Oxford

Library of Congress Cataloging-in-Publication Data

Tracy, Christine M.
The newsphere: understanding the news and information environment /
Christine M. Tracy.
p. cm.
Includes bibliographical references and index.
1. Journalism—United States—History—21st century.
2. Reporters and reporting—United States—History—21st century.
3. News audiences—United States. I. Title.
PN4867.2.T83   071'.300905—dc23   2012000744
ISBN 978-1-4331-1042-9 (hardcover)
ISBN 978-1-4331-1043-6 (paperback)
ISBN 978-1-4539-0592-0 (e-book)

Bibliographic information published by **Die Deutsche Nationalbibliothek**.
Die Deutsche Nationalbibliothek lists this publication in the "Deutsche
Nationalbibliografie"; detailed bibliographic data is available
on the Internet at http://dnb.d-nb.de/.

The paper in this book meets the guidelines for permanence and durability
of the Committee on Production Guidelines for Book Longevity
of the Council of Library Resources.

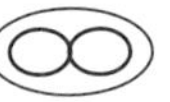

to John

CONTENTS

Contents

ACKNOWLEDGMENTS

I am truly grateful to all those who have inspired me to write this book and to those who supported its creation, development, and production.

I am especially indebted to Rensselaer's Dr. S. Michael Halloran, who saw Teilhardian studies as "a telos worthy of investigation" and encouraged me to embark on this less traveled scholarly road that is now my passion.

Media ecology is my intellectual home, and I am grateful to my friends and fellow media ecologists especially Lance Strate, who helped me vision the concept of an ecology of news.

I want to recognize the financial support and encouragement I received from the students, faculty, and administration of Eastern Michigan University, including my colleagues in the Department of English Language and Literature; English Department Heads Russell Larson, Rebecca Sipe, Laura George, Joseph Csicsila, and Mary Ramsey; Thomas Venner, dean of the College of Arts and Science; and Jack Kay, former provost. The difference between wanting to write a book and actually doing it is often time; I have received that gift from EMU and I am truly grateful.

Many eyes read and guided the evolution of this manuscript, and I am thankful to Amy Butters, Elizabeth Kirchen, Mekie Kukan, Debbie Marion, and Joel Seguine for their contributions to the manuscript's quality and professionalism. I also appreciate all the technical support I received from Jordan Brown.

My muses are many. I draw profound inspiration and insight from Monhegan Island and the community of artists, writers, and adventurers who are drawn to this powerful place.

A special thanks to the family and friends who offered encouragement and support especially my husband, John Strobel; my sons Brendan and Michael Lapham; Chris Adams, Corlis Carroll, Elizabeth Elvin, Patricia Fero, Katya Haskins, Tina Lincer, Anne Lopatto, Stacie Printon, Sue Quackenbush, Marilyn Ringer, and all the wonderful women I am blessed to know. I draw special strength from my late aunt and mentor Corinne C. Wojtala.

You are holding this book in your hands because of the expertise and guidance of Peter Lang's Mary Savigar. Thank you, Mary. I am truly grateful to you, to the reviewer, and to the production department at Peter Lang.

And finally Teilhard, my pilgrim from the future, I want to thank you for this amazing challenge. I have done my very best to give witness to your words and vision.

Namaste.

Christine M. Tracy
Ann Arbor, Michigan
December 17, 2011

INTRODUCING THE NEWSPHERE

I first started writing news when I was a junior at Mother Seton Regional High School for girls in Clark, New Jersey. I was the editorial page editor of the *Setonnaire* and more than the mild celebrity, I enjoyed seeing my words in print and having a public place to think out loud and create news—that singularly powerful blend of events, ideas, thoughts, information, and yes, hopes and dreams, too. I was constantly in awe—and still am—of my friends, the visual thinkers: the cartoonists who cleverly and quickly sketched their ideas and powerfully captured the essence of an idea. "Words were all I had (to take your heart away…)," as the Bee Gees crooned, but even as a fledgling journalist I sensed the organic, heady yet paradoxical power of this uniquely ephemeral yet lasting form.

It's pretty amazing actually that each day, all over the world, hundreds of thousands of people who call themselves reporters get up and go out and find out what is happening and let the rest of us know. We wake up and look at our BlackBerries, our iPads, and laptops or turn on the radio or television sets and yes, some of us actually read a newspaper that we may still pluck from our front porch in the morning. There it is, tried and true—news, in all its glorious and not so glorious incarnations.

It's the stuff of our lives and rich fodder for movies and novels. It's an oft-criticized but much coveted profession. It's an exciting and evolving academic discipline. It's also THE cornerstone of any viable and working democracy. Without a quality flow of reputable news and information it is impossible to have a true representative government. History and the heavy hand of censorship both around the world and yes, here in the United States, too, have proven this over and over again. None of this is new (or news: yes, I'm aware of the irony here). But what is new and what has dramatically altered the news environment, which I will hereto refer to as our *newsphere*, is the sheer volume of messages carrying the important distinction and designation as "news" and the proliferation of methods, channels, and networks to deliver it.

In addition to a greater amount of both information and disinformation masquerading as news, and the creation of dynamic and ubiquitous systems used to deliver it, news stories also have a different life span now that they are free from the traditional constraints of print and electronic broadcasting forms. They circulate and evolve and, like our own DNA molecules, some survive and thrive. Some die a natural and welcome death, while others are fueled by powerful winds of consensus, as well as by darker forces, such as unprocessed and unconscious acceptance, and last well beyond their time. There is also a dramatic increase in the volume of individuals creating and distributing news. From well-intentioned bloggers, the new breed of citizen journalists, and seasoned reporters doing a better job with powerful new tools, to bombastic cable television news show hosts, angry and vengeful agitators, and the highly literate bored with too much time on their hands, our virtual and physical newsstands are bursting at the seams.

Learning how to sort through the newsphere's noise and the clutter, and particularly recognizing unprocessed and unconscious messages, is a key focus of this book. But the real work is a deeper and more integral understanding of the effect news has on you so you can design a more viable and enlightened relationship with and in the newsphere. Please carefully consider precisely what news is doing to you: Is it distracting you? Probably. Is it informing you and making you a better citizen? Probably. Is it making you angry? Sometimes. Is it shifting your focus away from things you can realistically change and instead focusing your attention on problems with no solutions and tragic events that drain your energy? Yes. You can immediately change that once you become aware of it! Determining the personal relevance, significance, and value of each bit and byte as well as the imagery that enters your consciousness is something only you can do. When you assume this vital, personal responsibility and channel your energy and attention in a more conscious way you not only improve the quality of your own life, you concurrently enhance our collective endeavors.

The News Road Less Traveled

It is an oft-elusive yet elegant and paradoxical quality of human existence: We crave connection with others while simultaneously reveling in our fierce independence.

Journalism historian Mitchell Stephens brilliantly describes this basic human need, the desire to ask and also to answer, "What's new?" as a "hunger for awareness":

> More than specific information on specific events, the great gift a system of news bestows on us is the confidence that we will learn about any particularly important or interesting events. The news is more than a category of information or a form of entertainment; it is an awareness; it provides a kind of security.[1]

How do we maintain a sound balance between our insides and our outsides—our independence and our interdependence—in the digital age of iPhones, iPads, and 24/7 deadlines? We look anew at the traditional journalistic quality of relevance. Long understood as one of journalism's bedrock principles—news must be relevant to the audience—relevance means something very different now that the news environment is networked and readers not only easily talk back to reporters but can become reporters themselves aka "citizen journalists."

I would venture that most of us are similar in thinking to my friend Melody McCormick, a retired reporter. We want to determine just the right balance of outlets, sources, media, and channels to create that fat and happy feeling of being just full enough of news. Here's a quick glimpse at Melody's optimal network: "a combo of internet, newspapers, radio sites, magazines, maybe club or church newsletters, and social sites I cobble together to make me well-informed and also entertained." Sarah Merion, a young Boston executive who studied abroad in Buenos Aires, has different but not incongruent needs. Sarah works to increase the quality of the information that she is receiving. "I've placed myself in a position to receive news, but it's divide and conquer. I need to decrease the quantity and increase quality if news is going to have real value in my life and in my world right now," says Sarah. She has tried multiple methods to design a viable news network, and she believes the only way to create, maintain, and ensure quality is to consciously limit news exposure. "Once you consciously decrease the quantity, quality should improve," she says.

The news is there—no need to lament the demise of the newspaper and the robust social practice of journalism that continue to draw some of this country's (and other countries') best and brightest. What we need to do now is stop complaining that news is lost and newspapers (and the journalism that supported them) are dead, and instead focus on creating, distributing, engaging, encouraging, recognizing, respecting, and supporting quality news in all its forms and guises. I respectfully disagree with the throngs of professionals and scholars such as Alex Jones, who's *Losing the News* is a melancholic and nostalgic lament for "the kind of news I knew." Instead, I align with the progressive journalism scholars and innovative thinkers—notably Howard Rheingold, Jay Rosen, and Jeff Jarvis—among others, who are building on the fundamental understanding that what we have come to know as news—its form, style, and substance—has been irrevocably changed because of the advent and proliferation of digital technologies. Quite simply, news is very different now because of the way it is created, distributed, and used. It is time now to work to make it better. As always, it starts and ends with us.

We Are All Connected (Even If We Don't Know It Yet)

I begin *The Newsphere: Understanding the News and Information Environment* by doing what journalists do best: I tell a story. On August 12, 2004, the Northeast power grid collapsed. After the lights came back on, the first question everyone asked was "why did this happen?" Hindsight now provides a reasonable answer: the decision, or more accurately, the indecision of an Ohio power plant operator to stay connected to the grid after his plant failed instead of "shedding load," an industry term for rationing power during peak usage, had profound and far-reaching consequences for millions of people.[2]

In addition to dramatically and powerfully illustrating our deep and real interconnectedness, this story also illustrates what happens when total faith and trust are put in technology. This is the reality of the newsphere in which we live. It demands a new, more sophisticated and active awareness, and while it is not necessary to experience the dramatic shock of a power failure to awaken us, it truly is time to pay much more careful and close attention to the effect news, and indeed all the information that we receive and process, has on us, our communities, and ultimately our collective consciousness. One of the biggest news stories of 2011 is the birth and activities of the international Occupy Movement, whose slogan "We Are the 99%" demonstrates how collective awareness and the power it generates can work to reform even seemingly insurmountable challenges such as the corrupting effect of money on politics.

Just as it was difficult for the control room operator in the power plant to really see and appreciate the reality of the situation, our day-to-day lives and the challenges of survival often mask our awareness of our deep and profound interconnectedness. Perhaps even more important, we are often clouded and confused about our power to effect change in our world and how to use that power effectively. This is a fundamental and profound challenge of human existence. The relationship of our internal and external universes (which ideally manifests as our outsides matching our insides and vice versa) is broadly termed *Integral Philosophy*, and has attracted the time and energy of some of the world's wisest and most talented philosophers and scientists including Ken Wilber, Henri-Louis Bergson, Alfred North Whitehead, Jean Gebser and of particular note and focus here, the Jesuit mystic and paleontologist Pierre Teilhard de Chardin.[3]

Teilhard (1881–1955), as he is affectionately known, coined the term *noosphere* to describe the location, structure, and evolution of our collective thoughts and knowledge. He believed the human condition could be improved through participation in the noosphere. He understood the organic unity of this web of thoughts and its beautiful, symmetrical evolution: it's really quite simple, we move together as we move alone. "The human is not the static center of the world, as was thought for so long," Teilhard says, "but the axis and arrow of evolution—which is much more beautiful."[4] Everything was one in Teilhard's heart and mind, a vision of the world he developed as a young child living amid the rocks and stones of the volcanic peaks of Auvergne, France, and the natural beauty of its forested preserves. He coined the term, "le Tout," or "the All," to describe the organic unity of spirit and matter, science and religion, the human and the divine.[5] He used what we all have—heart, hands, and head—to transcend the dualities of human existence, and he then devoted his life to sharing that wisdom with us in profound and often heroic ways.

For these, and numerous other reasons you will soon discover, Teilhard is the inspiration of this book. His intimate knowledge of the Earth and of men's souls was preserved in the charter of the United Nations, among other places, and more than a century since his death, he has a devoted Facebook following: As of this writing, about 2,000 people follow him on Facebook, and the number continues to grow.[6]

Each day we are challenged to find balance between internal forces—the free will and agency that make you "you" and me "me"—and external forces— the desire for connection and satiation of our "hunger for awareness." How do we achieve balance in the newsphere? How do we do what Teilhard did and transcend this duality? We move in a new way: We expand our awareness and navigate toward a nonjudgmental dialogue news style. Just as Appalachian Trail (AT) thru-hikers who must ultimately leave the sanguine forest for the highly mediated world, we may need to step away from our daily dose of news to better understand our own relationship to news and information. Jack Magullian is a New Zealand hiker who fasts from news for months while trekking the AT from Georgia to Maine. "Once you've achieved some degree of separation from normal media, your perspective changes," says Jack, an expatriate who has logged thousands of miles on the trail.

A news fast is just one way to wake up from the trance that unconscious participation in the newsphere creates. We need to wake up to the illusion of news and look closely at the historical and theoretical forces that precipitated the structural decay of the American news and information structure. From fake press conferences and the churnalism that turns real news into rubbish, to the misuse of power and our own delusion, the stories we unconsciously consume are often just smoke and mirrors, or worse, deliberate manipulation that includes our own self-deception. Building on the admirable, poetic, and powerful work of the journalist Bill Moyers, the reception theories of John Fiske, and cutting-edge research on news processing, I will demonstrate that an honest integration and appreciation of news are often elusive, and I will offer an analysis of one of the most popular stories of 2009, the capture of the *Maersk Alabama* in the Indian Ocean. The takeover of an American cargo ship and its captain by Somali pirates is an illustrative example of how news is dramatized, commercialized, and packaged so it's simply not news anymore. For example, Comedy Central's Jon Stewart and small outlets, such as the *San Francisco Chronicle,* got it right, but the mainstream networks, such as NBC, did not.

The discussion then moves from a qualitative exploration of news and its delivery systems to an exploration of the architecture and systemic structures that support the newsphere. In addition to explaining how news stories emerge and evolve, this section implodes the traditional understanding of bias and explains the new role of experts in the news environment. Above all, it challenges both creators and consumers of news, especially journalism students, to move beyond their current understanding of news, to see news as a network—open, dynamic, and participatory.

John Dewey, one of America's most celebrated advocates of democracy, believed that informed citizens didn't need to know "all the facts of the world," and instead he recommended possessing a flexible mind and character. Likewise, in his *Five Minds for the Future,* Howard Gardner outlines the role of discipline, creativity, respect, and ethics, and the skill of synthesizing information in the digital age. Finding good journalism is indeed an art now, and news consumers and producers face both quantitative and qualitative challenges. Why is it that most of us feel we are starving amid news aplenty? We can get news from all over the world 24/7 on our iPads, BlackBerries, cell phones, laptops, and through ever-growing social networks as well as from traditional newspapers, magazines, television, and radio; yet somehow we don't feel satisfied. I will develop the paradoxical relationship of attention and distraction and explain the need for journalists to capture people's attention in a balanced way.

News, as it now functions in the newsphere, is no longer a fixed story written by one reporter. News stories evolve in and through dynamic networks. Designing and successfully navigating these networks require creating and maintaining a quality flow of information to ensure that you will find out what you need to know. I will offer specific strategies for both designing your network using multiple platforms and media environments, and also for filtering the news and information you receive. Use a simple checklist as well as the insights and personal experiences of young, savvy users, such as Ross Johnson, and the emerging news literacy work done by journalism and communication scholars that include Neil Postman, W. James Potter, John McManus, and Howard Rheingold, and you will come to understand how to place yourself in a position to receive a regular flow of timely news and information.

These "news as a network" strategies build on and extend our historical understanding of the role of news and the more traditional terms associated with quality journalism and news literacy as provided by online sources, such as NewsTrust.net and others. They can also help you test news sources, prevent bias, deal with overwhelm, and recognize when your time and attention are being manipulated. These strategies are also flexible enough to evolve with ever more sophisticated and pervasive news generating technologies, and when combined with an internal shift in perspective, they promise to clear and satiate your news consumption palate.

ENDNOTES

1 Stephens. p. 12.
2 http://wiki.answers.com/Q/What_is_load_shedding_and_why_is_it_carried_out Retrieved January 24, 2011.
3 MacIntosh. p. 2.
4 HP. p. 7.
5 The two main biographies about Teilhard's life used here were Ursula King's and Amir Aczel's. Please see the working bibliography included here.
6 http://www.facebook.com/login/setashome.php?ref=home#!/pages/Pierre-Teilhard-de-Chardin/26296145687.

TEILHARD'S RULES FOR NAVIGATING THE NEWSPHERE

"By virtue of the world's fundamental unity, once a phenomenon has been clearly observed, even only at a single point, its value and roots are simultaneously present everywhere."
—Teilhard de Chardin, *The Human Phenomenon*

"I am a pilgrim from the future," Teilhard told his young friend Jean Houston, as they talked and walked around New York City's Central Park many Tuesdays and Thursdays from 1952 until the elder priest's death in the spring of 1955. "We need more specialists in spirit, and perhaps you will be one," Teilhard told young Jean. "That has really stood by me for literally the rest of my life," said Houston, who discovered the true identify of her friend, "Mr. Tayer," years later when she was an undergraduate at Barnard College. She began reading *The Phenomenon of Man* and realized its author was none other than the Jesuit mystic Pierre Teilhard de Chardin.[1] "I literally ran into him. He propelled me into the life I now lead," Houston said of her three-year relationship with Teilhard when she was a teenager living in New York City.[2]

Houston, a scholar, philosopher, and researcher in human capabilities, is an advisor to UNICEF, a consultant to the United Nation's Development Program, and one of the principal founders of the Human Potential Movement. "We are being called to a higher order," Teilhard told Houston as he described the noosphere, the word he invented to describe the collective mind of the planet. "The earth has grown a new skin, a weaving of the consciousness of the planet," Teilhard said. He described this skin around the earth as a "field of mind," a living membrane that would grow in "density and complexity" as it reflected the "irresistible tide of intelligence." We now see the accuracy and elegance of his vision as the growing global web of connections that sophisticated wireless computer networks and digital technologies make possible. These are "outer forms of this inner change," Houston said as she shared the insights of her mentor, friend, and surrogate father.

Teilhard encouraged Houston to "travel, travel, travel and learn, learn, learn," and that is what she did visiting more than a hundred countries and studying different living cultures. As a social artist, Houston believes the communal fabric of information, ideas, and experiences is dissolving the "membrane that keeps people from different cultures separate and insular" and promoting a greater flexibility of thought as well as the ability to recognize unconscious patterns. We are able to "exchange social DNA at a remarkable rate," says Houston, and she believes we have not been prepared for present times—a world of radical complexity, acceleration and emergence. Much of the increasing amplititude of conscious awareness is coming from the Internet, according to Houston, who beautifully describes the growing interconnectedness of human beings as a "cosmic dance."

I have taken the great liberty of stepping into young Jean Houston's shoes and imagining what it would be like to walk with and talk to her beloved "Mr. Tayer." If we were to ask him questions, as she did, about the human condition and our collective consciousness, how would he answer us? What advice would he give to "newspherians" attempting to navigate the sophisticated global news and information environment and the complex tools and systems that support it? How can we best function as "spiritual beings having a human experience"? Through a collation of the work of numerous Teilhardian scholars, integral theorists, a personal interview with Dr. Houston, and my own knowledge, vision and understanding, I have constructed the following rules for navigating the newsphere:

1. *We are all connected.*
2. *We evolve through our interconnectedness.*
3. *We satisfy our innate hunger for awareness when we consciously consume and create news and information.*

4. *We use technology to help us build and maintain connections.*
5. *We are often not aware of our interconnectedness.*
6. *We must "see or perish."*
7. *We can design news networks that sustain and support our interdependence.*
8. *We are called to build an integral form of journalism with ourselves at the center.*
9. *We are more powerful than we know.*
10. *We evolve the newsphere when we take action and balance action and acceptance, an integral worldview.*

We long for connection, a hunger for awareness of the other. We are evolving through our connections with each other, which are becoming more complex, powerful and sophisticated. Teilhard defined this as the noosphere, the glue that binds us together in a cosmic sense. It is an integral consciousness—the understanding that the human is the "axis and arrow of evolution." This web of connections is a source of often untapped energy. It is how we grow and change both individually and collectively. Technology helps us grow this global network of complex connections in dynamic and powerful ways.

However, despite our "hunger for awareness," the primary motivation to send and receive news, we are often deluded and unaware. "What stares us in the face is often the most difficult to perceive," wrote Teilhard. We must wake up, break through the illusory world of news that deludes us, and take responsibility for our own power and the deep dialogue that is possible. The role of perception is critical in the news environment. Teilhard believed that the whole of life lies in seeing— in learning to focus on what is essential and important and to see with more perfect eyes. *See or perish: this is the human condition.* "The history of the living world can be summarized as the elaboration of ever more perfect eyes within a cosmos in which there is always something more to be seen," says Teilhard We must focus our eyes correctly and recognize patterns.[3]

News stories have lost their essence—Teilhard's cosmic sense. Instead of connecting us to others, most news reports and coverage of events create a profound disconnect: these types of stories and this style of reporting (that we will define here as "debate-style news") makes us feel separate from the world instead of an integral part of it. This is not true of all news stories, as we will see. Throughout the history of journalism, there have been many great moments of honest truthtelling. But it is critical to receive a daily sustenance—to know which stories and outlets are draining our energy and which are uplifting reminders of a larger connection to the world.

Integral journalism features dialogue-style reporting and balances action and acceptance, our insides and outsides. News reporting, consumption, and creation can help build bridges and forge connections in new and powerful ways. When this is done, energy increases because it is easier to live without duality or judgment. Sharing information in a dialogue-based, integral way cleverly negotiates the paradoxical role and responsibilities of the personal and the universal mind. Humans are the "axis and arrow of evolution," according to Teilhard. Progress is made when each person takes responsibility for this power. He believed that it only takes one mind to change the world: "Truth has to appear only once, in a single mind, for it to be impossible for anything ever to prevent it from spreading universally and setting everything else ablaze."[4]

My favorite quote from Teilhard's writing is, "What would we do without our enemies?" They prompt action and movement in often uncomfortably ways. But by taking action, as the struggles of truth-tellers reveal, we discover our essence at the point of balance between desperate efforts to grow greater and the resistance to change.[5] This is how we balance our inside and our outside, the creative union that is our rightful inheritance and undertaking. The media ecology is different now: news is a network; open content is self-vetting; and the structure of the newsphere is a sophisticated heterarchy that requires an individual balancing of action and acceptance (an integral worldview) that produces a dialogue not debate style of news.

Pierre Teilhard de Chardin: A Pioneering Evolutionist

I was first introduced to the Jesuit paleontologist and philosopher Teilhard de Chardin in 1985 while working as a press aide for the Culinary Institute of America (CIA) in Hyde Park, New York. This famous culinary school was founded on the site of a former Jesuit seminary, St. Andrews-on-the-Hudson, where Teilhard and other Jesuits had been interned. Working in the press office involved giving visitors tours of the institute, and I soon discovered that visiting the famous Jesuit's grave was one of the highlights, particularly for international travelers. In fact, Teilhard's relatively obscure and very modest gravesite is regularly located and visited each day.

This early discovery initiated a lifelong interest in and commitment to better understand the life and work of this brilliant and enigmatic man. I next encountered Teilhard in the mid-90s when I was a graduate student at Rensselaer, a technological university in upstate New York. While at Rensselaer, I worked with other graduate students to produce *Computer-Mediated Communication Magazine*, one of the Internet's first "e-zines." Early adopters likened Teilhard's concept of the noosphere—a global brain or intelligence that enveloped the Earth—to the Internet and specifically, to the World Wide Web. I continue to find Teilhard's life and work powerful, inspiring, and critical to our understanding of the current news and information environment—our newsphere.

It has been more than 50 years since Teilhard's death, and knowledge of the writing, work and remarkable life of this accomplished Jesuit scholar continue to evolve. In many respects, Teilhard's vision for humanity—an organic unity created through recognition of a divine force behind both creation and evolution and an understanding of the role, responsibility, and power of individual human beings—still remains elusive. According to his biographers, Teilhard was dejected and discouraged at the end of his life when in late 1948, he failed to convince Pope Pius XII and other church authorities to allow him to publish the manuscript that would become his masterpiece, *Le Phénomène Humain (The Human Phenomenon)*.[6] The Jesuit order also forbade him from accepting a prestigious appointment at the Collège de France, a position Teilhard coveted.[7] At this time, Teilhard also grew distant from his longtime friend and companion, Lucille Swan.

Yet Teilhard never relinquished his ultimate vision: He never left the Church so he could freely publish his work, nor did he renounce his vows to marry Lucille. In fact, it is in remaining steadfast in his beliefs and commitments that his ferocious tenacity, ironclad will, and intelligence are most evident. "What would we do without our enemies?" he asks; and it is precisely here that he gives witness to his belief in the organic processes of life and their evolutionary nudge: "Exterior emergencies or shocks are indispensable to force individuals out of their natural laziness and set routines—and also to periodically break the collective frameworks that imprison them."[8] Sometimes those "exterior emergencies or shocks" are literally just that—a hurricane, tsunami, earthquake or an extended blackout. More often they are more subtle and personal, such as a job loss, heartbreak, or a chronic illness. Teilhard suggests moving with these challenges, and more than that, to not only see their natural place in the order of things, but to actually welcome these evolutionary challenges: It is how he lived his life.

Teilhard's belief in cosmic convergence is all the more poignant because he lived a life of personal evolution: He was the ultimate boundary pusher. One can only imagine how different his life story might have been had he renounced his vows and left the Society of Jesus. Yet he never perceived leaving as an option because he believed his relationship with the Church encompassed his relationship with his God. Thus he ferociously worked within the system he knew at great personal cost. Quite simply, leaving the Jesuits would have been the ultimate breach of faith, and Teilhard was the truest of believers.

Teilhard's lofty and yet unrequited lifelong goal was to reconcile science and religion.[9] Teilhard's work uncovering the fossils of the Peking man, the first known *Homo erectus* and a vital missing link between apes and humans, distinguished him as one of the leading scientists of his time: His diligence, tenacity, and forbearance helped popularize and disseminate Darwin's theory of evolution, which begged for proof. Teilhard was also a devoted member of the Society of Jesus, an intellectually rigorous and hierarchical order within the Catholic Church, whose leaders firmly held a literal interpretation of the Bible and its teachings on original sin and creation. As a child living in Auvergne, France, Teilhard conceived a philosophy of unification he initially called "le Tout" or "the All," which was drawn both from his own observations and from a reading of Henri Bergson's *Creative Evolution*. He believed in an "evolutionary force that brought constant change to living things," and everything evolved toward a greater complexity and spiritual unity—an Omega Point—and he spent his life trying to convince others of this truth.[10]

Key to "joining" this emerging collective intelligence, which Teilhard defined as the noosphere, is abandoning judgment. Gratefully, this is a process that unfolds most often over a lifetime but it is also a skill that can be actively practiced. Teilhard's gift was his organic understanding of the unification of all beings. "See or perish," Teilhard advises.[11] This brief admonition sums up his lived experience. To see as Teilhard saw—a World War I stretcher-bearer, a devout Jesuit priest, a tireless scholar and archeologist—meant constantly searching for a unified reality. One cannot fully integrate knowledge or possess ontological wisdom without having first lived an experience: the warrior must witness death, the priest must recite his prayers, the paleontologist must dig in the dirt—all activities in the physical realm, to actualize and access the spiritual realm. This is the lived experience, the life to which Teilhard gave witness. It is man as "the axis and arrow of evolution."[12] Teilhard carefully distinguishes between seeing and perceiving, and he acknowledges that the human drama of war heightened his senses:

> Undoubtedly it was during my wartime experience that made me aware of this still relatively rare faculty of perceiving without actually seeing, the reality and organicity of large collectives, and developed it in me as an extra sense.[13]

Perception is the foundation of Teilhard's cosmology, his grounding of the human experience, and the premise upon which he builds his explanation of the evolutionary process. In his introduction to *The Human Phenomenon*, Teilhard states, "The whole of life lies in seeing," and he attributes the forward movement of history to "the elaboration of ever more perfect eyes." When perception is the cornerstone of the human condition, then the continual refinement of one's vision is key to human evolution and survival. Expanded and expansive vision is "the mysterious gift of existence," according to Teilhard, who believed that it was "always possible to discern more"; thus "to see more and to see better is not, therefore, just a fantasy, curiosity, or a luxury."[14] This line of thought places human agency at the center of the evolutionary process and humanity as the key to unlocking the mysteries of the universe.

Teilhard was greatly influenced by the writing and thinking of Charles Darwin and Henri Bergson. From Darwin, he appropriated the concepts of time and change, common descent, natural selection, and survival of the fittest. From Bergson, he embraced the notion that human evolution is creative and paradoxically and simultaneously random yet purposeful. In *The Human Phenomenon*, Teilhard identifies human beings "in their totality" as the key to understanding the evolutionary process, which, he states, occurred in four stages: pre-life, life, thought, and survival. He believed an evolutionary force brought constant change to living things, and he attempted to explain how this works on both micro (individual) and macro (global) levels. Teilhard saw the universe as a single organism, with an inside and outside, subject to two kinds of forces: "the measurable force of physics and a force residing in human thought." The future of evolution, according to Teilhard, was the Omega Point, "the ultimate site of convergence through evolution."[15]

Teilhard's personal goals and his intentions for his fellow scientists and philosophers—indeed for all humans—were highly ambitious. He believed that a true understanding and representation of the human condition (his human phenomenon) demanded an integration of the physical and the spiritual dimensions, which had heretofore been separated. He identified the place of spirit within the physical realm and called for a closer study of the relationship of spirit and matter:

> The time has come for us to realize that to be satisfactory, any interpretation of the universe, even a positivistic one, must cover the inside as well as the outside of things— spirit as well as matter. True physics is that which will someday succeed in integrating the totality of the human being into a coherent representation of the world.[16]

Throughout *The Human Phenomenon*, a poetic yet detailed summary of his cosmology, Teilhard identifies two different kinds of energies, tangential and radial, and he describes how they work to change the world. "We certainly feel the two opposing forces combine in our concrete acts," says Teilhard, but he readily admits that fully understanding the spiritual power of matter is "the most difficult of all readings to perform."[17] The challenge of this task, however, is neither beyond him nor beyond the scope and complexity of his writings, and his explanation of how spiritual forces work both individually and collectively forms the foundation of his ideas about unification, the noosphere (collective consciousness), and evolution. Teilhard describes this spiritual energy in a voice that echoes his reverence for both its power and for the knowledge of its existence in a prayer he composed on New Year's Day 1932 while traveling with members of the Yellow Cruise, an archeological expedition to China:

> Above and beyond us, a supreme energy exists, one that we must well recognize— because it is superior to us—the magnified equivalent of our intelligence and our will.… Of this universal Presence, which envelops us all, we first request to reunite us, as in a common living center, with those we love and who are starting, though far away from us, this new year.[18]

It is this spiritual energy, the combination of both radial and tangential energies, that works in concert to both create and elevate consciousness: This is the site of evolution and a close approximation of how spirit and matter combine. It is the union of tangential (outside or matter energies) with radial (inside or spirit energies) that forms the collective consciousness and explains how it evolves. Teilhard termed this union and this space the "noosphere."

As a philosopher and phenomenologist, Teilhard invented the word "noosphere" to describe the layer of thought and its corresponding energies that surrounded the physical biosphere of the Earth. As a scientist, Teilhard worked to understand the evolutionary process on a physical or biological level. He believed that the Earth was a living organism that evolved through the progression of human consciousness and that the noosphere contained humanity's collective ideas, interactions, and energy. All humans contribute to this existential sphere and it is key to the future of the planet. According to Teilhard, collective and reflective thought will ideally converge at an Omega Point as individuals grow in consciousness and become more human. Teilhard termed this transformational process "hominization" and its conclusion "Omega," the last letter of the Greek alphabet.

In *The Human Phenomenon*, Teilhard thus describes the location and constitution of the noosphere:

> Above the animal biosphere there is a human sphere, the sphere of reflection, conscious invention, and felt union of minds (the noosphere, so to speak) and at the origin of this new entity...there is a phenomenon of special transformation which affects preexistent life: hominization.[19]

There is much speculation among scholars, writers, and thinkers that the current Internet, the worldwide network of computers, is "the mechanical apparatus of the noosphere." Teilhard also referred to networks in his explanation of how the noosphere functioned—"the Noosphere—is multiplying its internal fibers and tightening its network and simultaneously its internal temperature is rising, and with this its psychic potential."[20] While the complexity and sustainability of Teilhard's thought are evidenced in these inquiries, equating the noosphere to the Internet limits Teilhard's vision to one technological development. This question may be better answered by viewing human activities and consequences of digital technologies, such as the Internet, as evolutionary examples of noospheric activities on the Earth. Teilhard believed that war and peace were in the "minds of men." Robert Mueller, a longtime high-ranking U.N. functionary, thought of the U.N. as a "noosphere." It follows, then, that the world of news—the newsphere as we exploring and understanding it here—may also be considered a noosphere—a layer of thought enveloping the world.

The Collapse of the Northeast Power
Grid Illustrates Interconnectedness

At approximately 4:11 EDT on August 14, 2003, 50 million people in the Midwest, northeastern United States and Canada were stranded in subways and elevators, sat in their cars because traffic lights went out, watched as their computer monitors and cell phones went black as life as most Americans and many Canadians know it came to an abrupt and uncomfortable halt. While generators provided essential services to hospitals, and television and radio stations remained on the air, telephone, transportation and most essential services—including the availability of water—were either interrupted or stopped completely. It was the ultimate eco-jam—no more cell phone, no more car, no more elevator, no more lights, no more water—you're literally and figuratively off the grid.[21]

And, as most crises do, the August 2003 blackout brought out both the best and the worst in people. New Yorkers and those who were in New York City during the September 11, 2001, destruction of the World Trade Center were particularly distressed when the early effects of the blackout—no power, chaos, and masses of people in the streets—mimicked 9-11. "I was one month into my internship at St. Vincent's hospital when the blackout happened," a young health care worker, AMC, wrote in a blog post. "I remember the lights at the nursing station going out and hearing the emergency generators come on. Most of the staff, who had worked through 9-11, were immediately afraid this was another terrorist attack, and there was palpable relief when the radio reported the cause of the outage."[22]

For Zac, a New York City native and *New York Times* blogger, it was a "fantastic experience" that he hopes for again. Many New Yorkers joined rooftop parties where they barbecued and hurriedly consumed soon-to-spoil desserts and ice cream and drank beer before it got warm. "It made you notice things about New York that you'd never get the opportunity to otherwise appreciate (crickets! moonlight lighting the streets!). Despite the obvious inconveniences it caused, I think it is important that we are reminded every once in awhile how much we are dependent on technology. Ever since, I have wished for an intentional 24-hour blackout once a year," writes Zac.

As Zac so wisely noted, the collapse of the Northeast's grid and the power failure it caused provided a powerful reminder of the mostly unconscious reliance upon technology and external energy sources. It was, as Marshall McLuhan writes in *Understanding Media*, a true collision of forms—the moment "of freedom and release from the ordinary trance and numbness imposed by them on our senses."[23] So while some gazed at the stars above Toronto and Manhattan for the first time and the lights came on again for the millions left in momentary darkness, the focus quickly shifted to discovering the root cause of the massive and debilitating outage and ideally preventing its recurrence.

What Happened? Who Was to Blame?

August 14, 2003, was an unusually hot day (88 degrees Fahrenheit for most of the effected areas),[24] so electricity use was at its peak and the fragile system was under great stress from air conditioners and fans. Unlike other power-generating systems, such as coal or natural gas, electricity is difficult to store and there is literally nowhere for the power to go once it is generated. The deregulation of the electric power industry encouraged the development of power-generating places in geographically remote distances from the customers who actually use the electricity. As a result, electrical power often has to travel great distances from the place it is generated to the place where it is used.

This dynamic and complex flow is amazingly and routinely managed by control room operators through a network of circuit breakers. All the power that is generated must be sent somewhere immediately, so the control room operators work to ensure a regular flow of power by switching transmission lines and generators to prevent spikes, which cause circuit breakers to trip. Their failsafe backup option is termed "shedding load," which means they cut a specific amount from the larger grid to prevent the collapse of the entire system. "Every major blackout that has occurred has occurred because someone somewhere faced the decision of whether or not to shed load and hesitated," writes software engineering expert Bret Petticord.[25]

As millions of Canadians and Americans struggled to get home from work, find food and water, cool off and otherwise cope without the seemingly unlimited power to which we have become accustomed, early reports of the cause of the breakdown surfaced. Like two boys caught in a playground fight, Canadian Prime Minister Jean Chrétien and New York Governor Pataki blamed each other: Chretien attributed the collapse of the grid that left his countrymen in the dark to a lightning strike in a power plant in upstate New York. Meanwhile, Pataki claimed the massive outage did not originate in his state but "west of Ontario" where it "cascaded from there to Ontario, Canada, and throughout the Northeast."[26] Neither was accurate.

Hindsight, as they say, is 20/20, and evidence now supports the finding[27] that the first critical event occurred when First Energy, an Ohio utility company, had a silent failure with its alarm system.[28] At 2:14 p.m. EST on August 14, 2003, First Energy stopped getting alarms, which is not unusual and at this point, there was no reason for intervention. However, system operators rely heavily on audible and on-screen alarms, and their malfunction is a serious, if not critical, problem. At 2:41 p.m., the computer server hosting the alarm software shut down and the First Energy's IT staff were alerted. At this point, both the alarm system and the system designed to back it up failed. First Energy's IT staff then restarted the computer server that shut down the alarm system and believed that they solved the problem and now had a functional system. But the alarm system was still not working, and system operators had not been alerted and informed of the malfunction.

Meanwhile, the power grid was starting to get jammed up. Transmission lines were getting tripped and shutting down—the first clues that something was really wrong. First Energy System operators started to get one, two, three calls—call after call after call—from operators at other power plants warning them of the potential danger. Technical writing consultant Mary Aiello closely examined the communication of the operators and discovered that at 3:35 p.m. First Energy's control room operators began receiving phone calls informing them that the spikes on transformers were triggering "over-excitation" alarms indicating that there were "radical and dangerous fluctuations in grid activity."[29]

In his third and final call, a Perry Nuclear Plant Operator in Cleveland, Ohio, told First Energy operators that his meters were "still bouncing around pretty good" and "I know something ain't right."[30] First Energy operators examined their monitors and alarm system (which were not functioning and displaying old data at that time). This was their unfortunate reply to the early warning: "It's got to be in distribution…or somebody else's problem…I'm not showing anything."[31]

An hour passed, additional phone calls came in reporting tripped and overloaded lines, and the First Energy control room operator *finally* began to recognize that there was a problem in the system—his system. Yet despite this early warning and what was a quintessentially karotic moment—a time for quick and decisive action—the First Energy operator did nothing: He did not inform other operators (encouraging them to quickly shed load) or take action himself (remove this operation from the network). As we now know, and the "Final Report on the August 14, 2003 Blackout prepared by the United States and Canada" confirms, this decision (or perhaps his indecision) was critical. Engineers and other experts who investigated the grid failure concluded that the "cascading blackout could have been prevented if the systems operator(s) at First Energy had shed most of the load from the Cleveland-Akron area."[32]

Why It Is Important to Learn to Trust Our Own Instincts

While the authors of the Blackout report acknowledge that "it is not humanly possible for one person to understand all these events simultaneously," and there was no punitive reaction to the report's conclusions and findings, there is culpability. It is very easy to forget or perhaps refuse to acknowledge that doing nothing is also a choice—a choice often grounded in the denial of our reality. What should be done when one is faced with mixed, conflicted and contradictory information? What strategies work when one must decide on an appropriate course of action when, like the control room operators, we get information that conflicts with what we see before us? Whom should we trust? Is this what happens when trust is placed in machines and monitors instead of human beings?

These are difficult and timeless questions cast in a new light in the digital age of increasing global interdependence supplied by complex information networks and power structures. Yet we must ask and seriously consider them because, as we can see, the consequences are far-reaching and often irrevocable. Perhaps the most primary is the pressing moral imperative inherent in message reception—

on any level. "At the end of the day," Mary Aiello writes, "in a world involving (almost constant) interactions between man and machine, there is a quiet voice that should remind us…(to) sustain compassion, active listening, and human rationality." Bill Moyers, one of the country's premier journalists, likens this still, small voice within to a personal First Amendment, that "would protect the quite fragile voice that occasionally rises uninvited to say, "That's not so: That's not the truth.""[33]

Like the First Energy operators, intuition cries "something ain't right," yet despite this clear warning, the real truth is shut down and shut out. At first glance, this appears to be a parable about technological dependence and the dire consequences of Technopoly—a culture and state-of-mind devoid of values —that Neil Postman, the founder of the field of Media Ecology, warned about. It is also nature's ultimate "eco-jam," and more powerfully than the obstructionist antics of cultural jammers, it dramatically and urgently issues a call to wake up from the illusory trance—the daydream that ignores increasing dependence upon external energy sources and mindless devotion to technology. This is what happens both individually and collectively when we stop trusting ourselves and each other and abdicate the very real responsibility for recognizing, speaking, and sharing the truth. If there is any doubt that we are all connected, the consequences surrounding the ultimate and almost total failure of the Northeast power grid should remove any doubt of the paradox of our interdependence: We are only as strong as our weakest link.

Who Decides What Is News in Newsphere?

Much like the Northeast grid that supports and reinforces our energy dependence and interconnectedness, most of us are likewise dependent on the *newsphere*—the 24/7 global news and information delivery network. From Twitter and Facebook mini-feeds to the Associated Press and CNN, the media landscape is constantly evolving. Previously, news moved from producers to consumers in a primarily unidirectional way—newspaper publishers, editors and reporters decided what was news and what was newsworthy, and they created a product—the daily newspaper—to deliver news to the citizen consumer. The unidirectional legacy news model is being replaced by a much more fluid, flexible schema, which finds citizens producing as well as consuming news and information. In this two-way digital media environment, the creation and consumption of news is best viewed as an ecosystem. Much like the First Energy engineers faced with the power to choose whether or not to shed load and cut off their region from the grid, individuals are now both producers and consumers of news.

The short-term consequences of this conscious exercising of news judgment by a more enlightened audience now fluent in the language of the two-way digital environment are shifts in agenda-setting, framing, story selection and numerous others. The long-term consequences—yet to be seen—are profound. As individuals consciously and carefully consider qualities of news and their judgments are duly recorded and quantified, both what constitutes news and the location of responsibility for the exercise of news judgment will shift. Quite simply, who decides what is news will change and in fact, is changing right now. Are citizens ready to assume the personal responsibility for exercising news judgment that historically belonged to publishers, editors and professional reporters? The answer is emerging.

News Creation and Distribution Is an Ongoing Process

The newsphere places novel yet real and pressing demands on the news consumer now challenged to learn a new version of news literacy to filter the news noise polluting the world of journalism. Before the advent of digital technologies and the many-to-many communicative forms they created, news consumers relied on a hierarchical and industrial model of news, which predominantly relied on analog technology to create print and electronic broadcast media. The qualities of that news included centralization, filtering, one-to-many distribution, and profitability. In contrast, an ecological model of news and information delivery uses digital technologies (computer networks supported by the Internet's World Wide Web primarily) to produce news and information that is decentralized, unfiltered, many-to-many, and egalitarian. This powerful shift is dramatically changing the newsphere and all the news within it.

Shifts and trends in communicative transactions include user-generated content, which in the journalism discipline is often described as the citizen journalism movement. In describing these discourses as "voluntary" and "more beneficial," Steven Cooper, writing in his *Watching the Watchdog: Bloggers as the Fifth Estate*, has identified a critical component of the ecological model: the value of individual responsibility in the process. Much like natural selection in the biological world, the internally motivated drive to both search for – consume—as well as create—produce—meaningful and relevant news and information is the crux of the emerging collaborative model. Even more to the point, Cooper describes the ideal situation as Darwinian: "…the fittest ideas prevail because they are based on the strongest arguments, which are the arguments most persuasive, and hence most acceptable to the participants."[34]

The Newsphere: Understanding the News and Information Environment introduces, develops, explores and evolves a new and truly ecological definition of news:

> When viewed as an ecology, news is not a product to be consumed but a conscious act to engage with and produce shared information that has value in a community: This is how cultures and societies create their histories.

In this systemic cultural context, news is no longer viewed as merely a hierarchically derived economic transaction but as a progressive and egalitarian opportunity, responsibility and challenge for citizens to create and build community through genuine dialogue. A simultaneous focus on both consumption and production—a genuine ecological approach to news—allows us to open up public discourse to the collective level in new ways. When viewed as a system and an ecology, consumption of news becomes a conscious choice involving knowledge generation, information creation, and public distribution. It requires the audience to question and it shifts the location of news judgment from the externally oriented editor to the internally focused individual.

An ecological mode of news puts quality not quantity in the foreground. Consumption of news, as defined here, is a conscious choice necessitating informed thought. It sheds a different light on the traditional concept of news judgment. It allows the audience to question, and it requires its participation. By questioning traditional news judgment, audiences can set an alternative agenda and close the loop in the consumption-production components of this ecological approach.

Ich bin ein Media Ecologist

"Ich bin ein Media Ecologist," declared Fritjof Capra, physicist, systems theorist and author of the *Tao of Physics*, in a clever borrowing of President John Kennedy's phrases when Capra addressed 2008 Media Ecology Association conference-goes in Santa Clara, California. Capra is strikingly accurate because we are indeed all media ecologist (some of us simply don't know it yet!). Media ecology is an emerging and progressive field of study that elegantly yet practically weds a desire to optimize the power of ever-evolving communicative tools with a deep and profound humanism.

A favorite definition of media ecology comes from a group of young filmmakers visiting Ann Arbor during its annual film festival. They designated themselves as both media ecologists and cultural jammers, so during the question-and-answer session, when asked for their definition of media ecology, "pattern recognition" was the prompt and immediate response. Media ecologist first recognize the need to make sense of complicated communication systems and second to learn how to better use them. It's simple yet extraordinarily powerful—what tools are available (and what tools need to be acquired) to help make sense of the often chaotic and disorienting mix of ideas, information, descriptions, assumptions, and yes, that thing called NEWS, that constitute the ever-evolving and morphing media worlds?

Media ecology as a field of study originated with the late Neil Postman. Postman is most popularly known for his 1985 *Amusing Ourselves to Death*, a critical perspective on cultural impact of television. When Postman, public intellectual, media theorist and cultural critic spoke at the first convention of the Media Ecology Association in 2000, he firmly established the discipline's humanistic roots and goals: "…as I understand the whole point of media ecology, it exists to further our insights into how we stand as human beings, how we are doing morally in the journey we are taking."[35] Postman founded the field of Media Ecology and inaugurated it as a formal academic discipline when he launched the program of study at New York University.

The etymology of the word ecology is "house," and Postman uses the word both metaphorically and practically to urge us to keep our earthly house in order. Writing in his *Amusing Ourselves to Death*, he explains, "We use the word ecology to suggest that we were not simply interested in media, but in the ways in which the interaction between media and human beings gives a culture its character; and one might say, help a culture to maintain symbolic balance. If we wish to connect the ancient meaning with the modern, we might say that the word suggests that we need to keep our planetary household in order."[36]

In its essence, media ecology looks at a culture in the biological sense of the word; it is a communication theory based on scientific and biological metaphors, writes Marc Leverette, an award-winning American photographer, author, and multi-media artist.[37] In biology, if something new enters a culture, it changes the entire culture, not just the entering phenomenon. This is an underlying principle firmly rooted in both systems theory and ecology. "When a new factor is added to an old environment, we do not get the old environment plus the new factor, we get a new environment," says communications professor Josh Meyrowitz, author of *No Sense of Place: The Impact of Electronic Media on Social Behaviour.* Thus, the environment is always more than simply the sum of its parts.[38] This perspective and form of examination is the work of the both the traditional and media ecologist.[39]

The concept of media as environments is one of the key tenants of the media ecology tradition. From this perspective, media "become, and are our culture because of their pervasive effect on everything we do."[40] So instead of fixed forms that simply send and receive information or more disturbing, tools to release poisonous fare on an uniformed public, media ecologists see media as dynamic environments and systems that evolved (and continue to evolve) as a result of the way humans use and respond to them. Media ecologists work to better understand this power and creative potential and the corresponding responsibility inherent in using communicative tools, practices and institutions wisely and well. Years before the social media phenomenon, Postman asked: "To what extent do new media enhance or diminish our moral sense, our capacity for goodness?"[41] Answering this question individually and collectively remains a societal priority.

These are complex challenges for which Pulitzer Prize–winning reporter Alex Jones and other experts in the journalism field have yet to discover workable solutions. Like William McKibben, Neil Postman poetically articulates the very real harms created by the American technopoly and the pervasive and dire societal consequences of the unexamined media reception, but he does not offer to "fix" the problem. The discussion turns now to a close look at the product and consequences of an unexamined news environment—the veil and illusion of news—and begins to frame a new and more evolved kind of journalism.

ENDNOTES

1 Teilhard's writing is vast, complex, and challenging to novice readers. A valuable introductory source is Demoulin, Jean Pierre. *Let Me Explain Pierre Teilhard de Chardin*. Harper & Row: New York, 1966.

2 Houston talked with me about her transformative relationship with Telhard de Chardin during a phone interview on May 4, 2011. The text also contains segments of her interview with Andrew Cohen of EnlightenNext magazine on Saturday, March 27, 2011, which was part of the magazine's "Awakening To Your Highest Self: Tales of Transformation from 30 Plus Spiritual Illuminaries" and is archived at http://www.enlightennext.org/webcasts/. Dr. Houston is a prolific author—26 books—and human potential ambassador—she has visited more than 100 countries to study and learn from them, as Teilhard recommended. Detailed biographical information about Houston is available at http://www.jeanhouston.org/ Downloaded March 29, 2011.

3 Demoulin p. 116.

4 Unpublished extracts from Le Christique (1955) Conclusion: The Promised Land.

5 Demoulin p. 127.

6 *HP.*

7 pp. 211–212.

8 HP p. 97.

9 King p. 36; Aczel p. 75.

10 Aczel p. 75.

11 HP p. 3.

12 HP p. 7.

13 HP p. 53.

14 See footnote 3 from previous text.

15 Aczel pp. 189–190.

16 HP p. 6.

17 HP p. 29.

18 Aczel p. 165.

19 HP p. 256.

20 Qtd. in King p. 89.

21 The eco-jam is much like its kindred spirit, the cultural jam, a deliberate social statement aimed at disrupting the unconscious thought process of media messages—particularly advertising.

22 AMC's post is from a August 7, 2008 *New York Times* retrospective piece, "The Blackout, Five Years After," *New York Times* Downloaded September 29, 2010 from http://cityroom. blogs.nytimes.com/2008/08/07/the-blackout-five-years-after/

23 McLuhan, *Understanding Media* p. 55.

24 Wikipedia entry "Blackout of 2003" downloaded September 24, 2010.

25 Brett Petticord wrote this explanation on his April 13, 2004, blog post "How Did the Blackout Happen?" in a piece titled "Testing Hotlist Update." Downloaded September 24, 2010.

26 Brett Petticord wrote this explanation on his April 13, 2004, blog post "How Did the Blackout Happen?" in a piece titled "Testing Hotlist Update." Downloaded September 24, 2010.

27 See the "Final Report on the August 14, 2003 Blackout."

28 Petticord p. 1.

29 "Final Report on the August 14, 2003 Blackout" p. 65. ("Investigating the Blackout of August 14, 2003: Voices Left Behind in the Darkness." Master's thesis April 29, 2008).

30 "Final Report on the August 14, 2003 Blackout" p. 65.

31 Ibid.

32 Petticord's blog p. 4.

33 Moyers Illusions of News video.

34 Cooper p. 279.

35 Postman cited in Lum p. 69.

36 Postman, *Amusing Ourselves to Death* p. 62.

37 Please see Leverette's essay "Toward an Ecology of Understanding: Semiotics, Medium Theory, and the Uses of Meaning," *Image & Narrative* Issue 6, January 2003.

38 Meyrowtiz, *No Sense* p. 19.

39 Ibid. See Leverette 2003 above.

40 Gencarelli cited in Lum p. 247.

41 Postman cited in Lum p. 67.

$\cdot\ 2\ \cdot$

THE ILLUSION OF NEWS

You shall no longer take things at second or third hand…
You shall not look through my eyes either, nor take things from me,
You shall listen to all sides and filter them for your self.
  —Walt Whitman (editor, Brooklyn Eagle), "Song of Myself"

On Tuesday, October 23, 2007, as wildfires raged throughout California, Harvey E. Johnson, FEMA's deputy administrator, held a press conference in Washington, DC, to update the nation's press corps on his agency's response to this national emergency. Johnson told the small group of reporters assembled, who had been given less than 20 minutes' notice of the event, that he was very happy with FEMA's response to the fires so far: "And so I think what you're really seeing here is the benefit of experience, the benefit of good leadership and the benefit of good partnership—none of which were present in Katrina," said Johnson.

In fact, what astute journalists soon revealed was that the press conference was staged. Standing behind a podium in front of television cameras, Johnson answered six questions from his colleagues, who were posing as reporters. This "error in judgment," as it was later described in a FEMA press release, was soon widely reported in a variety of news outlets. It is perhaps too glaringly apparent from these and other more popular stories, such as the outing of former CIA agent Valerie Plame, which was brilliantly reported by PBS's *Frontline* and later made into a 2010 movie, *Fair Game*, and the 2007 release of Jeremy Scahill's 400-plus-page impeccably researched *Blackwater: The Rise of the World's Most Powerful Mercenary Army* that the watchdog function of the nation's fourth estate is still in great demand.

The Media Is Not Journalism and Journalism Is Not the News

Serving as "an independent monitor of power" is the fifth of nine principles Bill Kovach and Tom Rosenstiel outline in their seminal *The Elements of Journalism*, one of the clearest statements about the purposes of American journalism today. "What is journalism for?" we may ask. "Journalism is for democracy," according to Kovach and Rosenstiel.[1] In his ritual theory of communication, the late James W. Carey, former CBS professor of International Journalism at Columbia and a stellar media ecologist, pioneered a dynamic, culturally, and technologically aware approach to news dissemination and delivery. Unlike more scientific, sterile, and objective viewpoints, Carey believed that newspapers were "a form of drama" and that news "was not pure information but a portrayal of the contending forces in the world." In his 1993 article "The Mass Media and Democracy: Between the Modern and the Postmodern," Carey reminds us that journalism and democracy are intertwined, historically variable, and greatly dependent upon the affordances of current communicative technologies.

> The media have changed decisively in the last 20 years, both as technologies and institutions. Yet democracy has changed also; the ends of political life have been reconceived in recent years. There is a widespread demand for less pro forma political representation, whether by the press or elected officials, and for more real participation.
>
> Yet these changes only signal that the meanings of democracy and communication are historically variable. The meaning of democracy changes over time because forms of communication with which to conduct politics change. The meaning of communication also changes over time depending on the central impulses and aspirations of democratic politics. Neither communication nor democracy is a transcendent concept: they do not exist outside history. The meaning of these terms varies with available media and with whatever concrete notions of democracy happen to be popular at any particular time.[2]

Key to this analysis is Carey's notion that both democracy and journalism are fluid and flexible concepts grounded in specific histories and times, and dependent upon available technologies. Significant social, economic, and cultural trends, such as technological innovation, globalization, and political unrest, have created a new media environment and a new model for news production and delivery. The legacy model, in which news was produced and consumed in a primarily unidirectional way, is now being replaced by a much more fluid, flexible schema, which finds citizens producing as well as consuming news and information. We now need to adjust what we have come to know as journalism and what it's supposed to do and also stop equating the practice of journalism in this country with the media industry: *Journalism is not "The Media."*

Media and cultural observers, and yes, journalists too, now have a whole host of descriptors for their profession and practice. In addition to "Internet Journalism," used by Wikipedia to define Matt Drudge's work, the term journalism is often preceded by descriptors such as advocacy, citizen, community, online, public, precision, video, and way new. There are adjectives used to describe the practice and function of journalism—participatory, investigative, or civic; adjectives used to describe media that deliver news—print, video, digital, online, broadcast, or print; adjectives used to describe the genres of journalism—sports, celebrity, science, and environmental; and adjectives used to point out the profession's flaws—yellow, ambush, gonzo, and gotcha.

Correspondingly, there are a host of descriptors to define current media outlets and practices including mainstream media, alternative media, independent media, social media, and, finally, emergent media. Describing media as emergent is particularly useful in this analysis because it provides a background for the evolutionary process whereby media come into existence. It also more accurately depicts the existing technological affordances, distribution channels, and communicative forms now being used to conceive, design, share, publish, and distribute news. Finally, given the variety of adjectives used to describe media, using the word emergent emphasizes their social constructive aspects, which provides an important perspective for defining news as an ecosystem.

Media Products and Journalistic Practices Are Not the Same

Twitter reports of the 2011 revolution in Egypt and other civic uprisings in northern Africa, and indeed all around the globe, offer tangible evidence of the constantly evolving newsphere, which is experiencing dramatic change since the advent of digital technologies in the early 1980s. Likewise the first images the world saw of the terrorist bombing of the London subway in the summer of 2005 came from survivors' cell phones. These grainy but powerful images were almost instantly broadcast to global audiences on television and the Internet. Change, innovation, and experimentation is so predominant now that it has become exceedingly difficult to distinguish the communicative form that delivers the news from the practice of journalism, which motivates individuals to both create and consume news. This is a vital distinction and worthy of serious deliberation because the results of both these endeavors—the product of the media and the product of the practice of journalism—are not the same.

Too often media is substituted for or used interchangeably with journalism, journalistic practices, and news and information delivery. The analysis that follows makes a clear distinction between media, defined as the communicative form in which the news is conveyed, and journalism, defined as the discipline, practice, and ethical and democratic responsibility for communicating news to an appropriate public. For example, mainstream media (MSM) is most often associated with the major television networks, the media conglomerates, and large publishers that produce and distribute most American newspapers. In most current popular usage, it also refers to the practices and procedures that produce those products. But they are not synonymous. The social practice of journalism is not the media industry. This is a critical distinction: If the organizations and institutions that currently produce a majority of the news product consumed are structurally flawed, the practice of journalism is not.

This distinction is necessary for many reasons. First, it is important to distinguish the communicative form that delivers the news from the practice that created it. This will help to clarify the complex interrelationship of the two and further elucidate how both consumers and producers of these communicative forms manipulate and design them. Second, this approach may help rescue the profession and practice of journalism from its association with large media conglomerates, whose marketing agenda is degrading the quality of information that most mass audiences consider news. The analysis will first clearly define media and journalism, explain their complex and often symbiotic relationship, and then offer different language and a novel approach to explicate the current and future state of news generation and its consumption and use.

"There Are Media Everywhere.
There Just Isn't All That Much Journalism."

The goals and principles of journalism are concretized in the Project for Excellence in Journalism's (PEJ) statement of purpose:

> The central purpose of journalism is to provide citizens with accurate and reliable information they need to function in a free society. This encompasses myriad roles—helping define community, creating common language and common knowledge, identifying a community's goals, heroes and villains, and pushing people beyond complacency. This purpose also involves other requirements, such as being entertaining, serving as a watchdog and offering voice to the voiceless.[3]

In addition to this definition, PEJ also lists nine journalistic principles. These include an obligation to report the truth; loyalty to citizens; dedication to verification; independence; monitoring of power; providing a forum for citizen debate and discussion; striving to be interesting and relevant; keeping the news proportional and maintaining comprehensive reporting; and finally, exercising a personal conscience—a moral compass directed by ethics and responsibility.[4]

These are things journalists are expected to do and to believe. They are the practice, process, and the principles of the profession. They do not include profit margins, conglomerates, buy-outs and by-ins, the industry, the media, and the media monopoly. This is a very important distinction: The practice and profession of journalism is not the business of delivering news, a product generated by "The Media" and "sold" to national, international, and global audiences. Journalism must be defined separately from the news industry if the role of citizen journalists is to be recognized and seriously respected.

One of the best ways to discover the differences between the news that the industry packages and delivers to audiences and the news that contains enduring qualities and standards similar to those that the PEJ outlined above is to follow a breaking story as it is reported on television and to follow the same story as it is reported via social media and the alternative press. For example, CNN reported early on the morning of February 4, 2011, that American journalists were being attacked and there were no new images of the revolution in Egypt. Meanwhile, Nevine Zaki of Cairo shot and distributed a photo of Christians protecting Muslims as they knelt for prayer, which was immediately published on OpEdNews.com.[5] The photo caption read: "To those who suggest that the Egyptian revolution is a threat, that radical Muslims will take over and endanger Israel, here is an image that contradicts that negative energy. It has already been viewed over 120,000 times on one site alone." As I write this, ABC news has contacted Zaki and asked for permission to use her photo on its network television programs.

Engaged and tech-savvy citizens, such as Cairo's Zaki, are building a dynamic human network supported by ever more powerful and accessible digital technologies. They are demonstrating that journalism is now a layering process: "Ordinary" people and professional reporters alike are weaving a real-time news fabric that allows readers to follow a story and make up their own minds about its significance and impact. In contrast, the news that the media industry packages and distributes is often fixed and says to audiences, "This is the way it is. Look. See this (and only this) and think no more!"

Two of the most insightful scholars to help define the role of journalism in society are Michael Schudson and James W. Carey. Schudson is a Columbia journalism professor and MacArthur award winner. Carey was also at Columbia before he died in 2006. In *The Sociology of News* (2003), Michael Schudson devotes an entire chapter to defining journalism and does not exhaust the topic. He concludes with this definition: "[Journalism] is information and commentary on contemporary affairs taken to be publicly important."[6] Similarly, in an opening day address to Columbia's journalism students, Carey clearly defines journalism, explicates its role, and distinguishes the field from the media industry:

> Like the novel to which it is at every historical point connected, Journalism converts valued experience into memory and record so it will not perish….Journalism takes its name from the French word for day. It is our daybook, our collective diary, which records our common life. That which goes unrecorded goes unpreserved except in the vanishing moment of our individual lives. Here you will study the practice of journalism. Not the media. Not the news business. Not the newspaper or the magazine or the television station but the practice of journalism. *There are media everywhere…there just isn't all that much journalism.*[7]

In defining and distinguishing the practice of journalism from the media industry, Carey can be seen also to frame the role of the citizen journalist. Given the recording and memory-making role of journalists, it follows that this function can be assumed, and has historically been assumed, by ordinary citizens. This was indeed the case for much of the history of American journalism until the 1830s, when distribution of news shifted from periodic journals sold by subscription to penny papers sold daily on street corners.[8]

A Brief Theoretical Look at the Construction of News

News is often equated with culture, and the kind of news a society consumes both determines and is determined by the nature of that particular society. In form and content, the newspaper has historically served as both a mirror and a repository of the culture and society it serves. Carey describes the newspaper as "a presentation of reality that gives life an overall form, order, and tone."[9] Newspapers and their myriad offspring, which now include iPad-only editions, are a vital part of the newsphere. Ideally, audiences, readers, and consumers process the news they receive and use that information to develop a better understanding of the world and an evolving point of view on reality. The flow of news and information people receive in large part determines their reality and enables them to think and act in new ways. Thus both individually and collectively, informed citizens act within a society and make both conscious and unconscious choices in response to a myriad of situations. It is the collective responses to these situations that eventually become the political, social, economic, technological, and historical imperatives that shape and determine the society. The dominance of the traditionally printed newspaper is waning and new genres are emerging to replace it, yet the power and impact of news and information delivery continues to grow and extend to the farthest reaches of the globe.

In general, the power of new communication technologies to exert social control, as Robert McChesney, Ben Bagdikian, and other media scholars and critics confirm, continues to grow exponentially due in part to the dramatic advances in the technology used to distribute and produce news. Harold Innis closely studied the paper industry in his native Canada and concluded, among other things, that the power of communication technologies to artificially manipulate the flow of information is the driving force behind economic and political imperatives. Michael Schudson believes that the power of the news media comes from their mutually constitutive quality, a creative intersection of evolutionary and revolutionary models. Because communicative media, as all writing technologies, exist first in the minds of individuals and then are evidenced in the world, they create a polarity or duality, which is actualized in the actions of individual agents responding to exigencies, according to Schudson.[10]

What Exactly Is a Medium
and What Is Its Role in the Newsphere?

For a long time, the media that have historically contained and delivered news and information, such as newspapers, magazines, and newsletters, were seen merely as transmission channels. This makes sense when you consider stone monuments or illuminated medieval manuscripts. However, all media—from stone monuments to iPads—possess a dynamic quality. The evolution of these forms is much more apparent now because message delivery is almost instantaneously both across town and around the world. As the enigmatic Marshall McLuhan phrase—"the medium is the message" —aptly illustrates, a medium is not a fixed entity but a fluid and flexible form—a dynamic choice reporters, writers, and producers make when they are designing and composing a news story. The historic 2011 revolution in Egypt beautifully demonstrated how digital media and the social networks they create seem to be unbounded by time and space. So what we are witnessing is the unprecedented and apparently ever-expanding access to global news and information. In the newsphere, news now takes many forms including oral, textual, visual, and audio, and is delivered by both ad hoc and more formal, structured networks. Learning how to monetize the digital forms of news and the networks that support its delivery remains a constant and critical challenge to the news industry.[11]

Who Sets the News Agenda? A Case Study of News Creation,
Consumption, and Delivery in Baltimore, Maryland

By Helen Nevius Adamopoulos

As Twitter, independent blogs, and other new forms of news media rise in prominence and older print forms struggle, many questions arise about the form and delivery of news. For example: Are new media really replacing old outlets such as newspapers as the population's primary source of information? Do these new outlets adequately make up for the shrinking production rates of traditional media organizations? Or, will the decline of television, print, and radio news leave the public less knowledgeable about the workings of the world around them? Researchers for the Pew Research Center's Project for Excellence in Journalism (PEJ) set out to answer these questions by closely studying the local media ecosystem in Baltimore, Md., for one week.[12]

The PEJ researchers monitored all local media outlets in the Baltimore metropolitan area from July 19 to 25, 2009. Their sample included newspapers, niche print publications, television news programs, radio news broadcasts, and new media (such as Twitter feeds and blogs). After analyzing the content these local news sources produced, the researchers discovered that although the media landscape in Baltimore expanded dramatically with the advent of newer media forms, traditional journalism still dominated the newsphere: Older media—particularly newspapers—still determined what the public learns about local day-to-day events.

The 2009 PEJ study found that traditional media produced 95 percent of the news stories that contained significant new information or details. Traditional media also tended to set the agenda for other outlets with the stories they chose to cover. The study also notes that older outlets, such as newspapers and television news programs, have become multi-platform and are now using the capabilities of companion websites and Twitter to publish breaking news. For example, Baltimore's newspapers, and television and radio stations, published nearly a third of their stories on new media platforms; almost half of the newspaper stories in the PEJ study were posted online rather than featured in print. A key conclusion of their report is the important observation that new media working independently from mainstream news sources most often quickly and effectively spread stories from other sources without adding new information.

What Methodology Did the PEJ Researchers Use?

In order to conduct this study, PEJ researchers first conducted a formal audit and identified all local media outlets that were active and producing local public affairs news. From these, the researchers then selected only those outlets whose content could be captured in print or digitally for inclusion in the study. The 2009 PEJ team then saved all the content produced by the selected media outlets during the week of the study. This involved collecting hard copies of the newspapers, capturing radio shows through online streaming, and digitally downloading television broadcasts from a subscription-based online media database. Websites were captured at varying frequencies (such as twice per day) depending on how often they were updated. After amassing all news produced during the week, the PEJ researchers then analyzed each story. First, they recorded basic information, such as the story's headline and publication date. They then studied variables, such as geographic focus, format, and origin (i.e., whether it was original reporting or picked up from another news source), content description, and the broad story topic.

Here's How the Baltimore Ecosystem Works

The 2009 PEJ study analyzed 53 different news outlets that regularly produced local content in Baltimore. These sources included the following: four local television stations, five general interest newspapers, two long-standing specialty papers, four general-interest websites, five local blogs, two commercial radio stations and two public radio stations. Of these outlets, 41 generated 715 different stories concerning local events for three days during the week of the study. The remaining 12 did not produce any local content at the time of the study.

Television newsrooms tended to produce more content than other media, averaging 73 stories per station. Newspapers came in second and produced 37 stories on average per outlet. Television stations also devoted the most coverage to local issues: 64 percent of 6 p.m. newscasts concerned local events. Researchers found a local focus in 53 percent of newspaper stories, 52 percent of talk radio segments and 85 percent of new media postings (although these items were generally brief and derivative). Overall, 23 percent of television stories focused on crime. Print outlets predominantly covered crime and government—17 percent and 15 percent respectively. Radio focused 19 percent of its stories on government, which was also the new media's favorite topic (20 percent of postings).

Government officials—primarily the police—spurred the coverage of 63 percent of the stories in the study. Editors and reporters initiated 14 percent, while 10 percent came from colleges and universities. Citizens initiated 12 percent of the stories, and the remaining 1 percent resulted from spontaneous events.

The researchers identified six larger narratives that occurred during the selected week based on individual news stories from various sources.

Researchers analyzed each storyline and gathered the following data:

Story 1: Governor Announces New Budget Cuts

The first of the six stories studied centered on Maryland Governor Martin O'Malley's announcement that he wanted to cut roughly $300 million from the state budget. Traditional news outlets drove the coverage; only six of the 69 stories published throughout the week about the cuts were from new niche media. Most of the information in the week's reporting about this story about the budget cuts came directly from the governor's office; only 16 news items contained significant new information. Web sources and local blogs essentially ignored the proposed budget cuts. Outlets such as the *Sun*, WBAL, and niche print publications generated all of these stories except for the stories traditional outlets posted online. Finally, the governor triggered 71 percent of the stories on the topic with actions such as public statements and press briefing. Only 7 percent of these stories involved journalists delving deeper into the implications of the cuts or discovering and creating new angles.

Story 2: Police Officers Wounded in Shoot-Out

On July 18, an apparently drugged Baltimore man shot and wounded two city police officers before he was wounded himself and detained. A reporter for the *Sun* first broke the news using his Twitter feed connected to the newspaper: He posted that an officer had been shot. The Baltimore Police Department Twitter feed followed, asserting that, in fact, two officers were shot. The *Sun* reporter and the police continued to provide updates on the story using Twitter. While tweets launched and updated this story, overall most of the coverage came from mainstream outlets. The *Sun* and four local television stations used new platforms—mainly websites—to break the story and provide new information. The *Sun* produced three of the 13 stories that contained new information; nine more came from television stations, which emphasized the emotional aspects of the story while introducing new angles.

Story 3: University of Maryland, Baltimore, Was Chosen to Develop the Swine Flu Vaccine

On July 22, the University of Maryland, Baltimore (UMB), announced that it had been selected to test a new swine flu vaccine. Newspapers provided the most thorough coverage of this event; the *Sun* ran a 1,265-word front-page story on the vaccine trial the day after UMB made its announcement. The *Maryland Daily Record*—a niche print newspaper—also broke the story. Overall, newspapers generated seven of the 19 news items on the vaccine trial. Local television stations' websites and broadcasts accounted for nine percent, and the rest came from local radio station sites. The study did not find any mention of the story in local blogs. Unfortunately, the coverage of the story suffered from the lack of any original enterprise reporting. The university's news release and press conference served as the source for most of the stories published in all outlets. Only three news items contained significant new information, while three more added minor new details.

Story 4: The Auctioning of the Senator Theatre

Also on July 22, the city of Baltimore held a sidewalk auction to sell the Senator Theatre, a historic movie house. The city ultimately "won" its own auction and bought the theatre for $810,000. However, the local news outlets failed to accurately predict the outcome. Although the *Sun* mentioned the possibility of the city winning its own auction, other outlets did not. Instead, these outlets speculated that a local college might buy the theatre (although this institution was not even a bidder). Only three of the 15 stories on the auction mentioned that the city might win. Traditional media used new outlets to report and distribute this story. For example, WBAL radio reported the result of the auction on Twitter, and the *Sun* posted a brief story online. Newspapers drove the coverage; all of the stories that contained new information came from outlets associated with print publications.

Story 5: Listening on Buses

The fifth of the six narratives the PEJ researchers followed included a story about the Maryland Transit Administration (MTA). This statewide organization wrote to Maryland's attorney general and requested a legal opinion on a proposed plan to add listening devices to the video recorders already installed on MTA buses. The attorney general posted the inquiry online, where a blogger spotted it and made it into a major story. MTA quickly dropped the idea.

This story beautifully illustrates the complex and evolving relationship between new and traditional media. Paul Gordon broke the story on the blog *Maryland Politics Watch*. The *Sun* picked up on the story from the blog, which increased the story's coverage. This is an excellent example of a now common trend in the new-sphere: A new or digital media outlet breaks the story but wide distribution and dissemination most often requires pickup by a mainstream outlet. Coverage of this story also demonstrated another trend: Despite many outlets publishing the same story, there may be no new information reported. For example, only the *Sun* and the *Maryland Politics Watch* provided significant new information; the other 22 outlets that addressed the story simply repeated already published facts. Many new media outlets also posted or published material from other sources without properly crediting the original outlets, a very real and important challenge to source credibility on many levels.

Story 6: Concerns About Juvenile Justice

Finally, the sixth story the PEJ researchers focused on was a composite of a variety of reporting about juvenile crimes during the study week. Many stories came together and formed a larger narrative about attempts by Maryland authorities to prevent and punish juvenile offenses. Once again, traditional news outlets drove the coverage and produced 68 of the 78 news items. Specifically, the *Baltimore Sun* and WBAL-TV broke most of the stories and established most of the narratives. As was evidenced in the coverage of the other stories, many outlets simply re-report a story and do not do any new or original reporting. Only six of 55 traditional news stories studied here contained significant new developments. WBAL produced three of these; the *Sun* generated two, and the remaining story came from WJZ-TV. Other outlets and news items mostly repeated others' reporting.

New Media Trends PEJ Identified

Based on their study of news creation and dissemination in Baltimore, the PEJ identified several trends in the reporting of new media outlets:

1. As journalists spread news quickly through digital means and often forgo doing their own reporting, official accounts such as press releases have become increasingly important as sources of information.
2. Many independent news websites publish work from other news sources without crediting the original authors. Online sources also occasionally post or link to outdated news stories, providing inaccurate information.
3. The vast majority of news the Baltimore public consumes contains no new information or original reporting, which is sometimes called "enterprise reporting."

For example, almost 85 percent of all the stories included in the PEJ did not contain any new developments or details. Therefore the approximately 15 percent that remained did offer fresh reporting but it came almost entirely from traditional media. General interest newspapers such as the *Baltimore Sun* generated 48 percent of the stories that contained new information. Local television stations produced 28 percent of those stories, and specialty print publications accounted for 13 percent. Seven percent came from radio stations (all from information posted on their websites), and only 4 percent of the enterprise reporting originated with new media outlets.

Although traditional media produced most of the stories, data on news production throughout the last decade show that Baltimore's local newspapers are producing fewer stories than they once did. In the entire year of 2009, for example, the *Sun* published almost one third (32 percent) fewer articles overall than in 1999. As the newsphere evolves and its networks and tools adjust, so too does the location of news judgment and the form of news stories. Before we look closely at its evolution, we will first study the de-evolution of news so we can be alert to its consequences.

De-Evolution in the Newsphere: The Reporting of the Somali Pirates' Capture of Navy Captain Richard Phillips

Reports of pirate attacks were among the first stories reported by American journalists more than 300 years ago. It is hard to believe, but journalists are still writing about pirates! In fact, the reporting about the hostage-taking of Somali pirates was one of the top stories of 2009.[13] The following analysis of *Dateline NBC's* reporting of the capture of Navy Captain Richard Phillips powerfully illustrates the very real dilemma of news readers who are hungry for the truth but often served a less satisfying fare.

"When you hear a splash in the water, you will know it's me." Richard Phillips, the former Boston taxi driver and now cargo ship captain, tried to assure his crew that he would soon be able to outwit his Somali captors.

And Phillips was true to his word. He waited for the pirates to fall asleep and valiantly dove overboard, but his escape plan failed, and he was soon taken again, bound, gagged, and held at gunpoint on a small, hot lifeboat. The pirates had their captain, but the crew of the *Maersk Alabama* still had the ship and all its resources. Within hours, two USS warships—the *Bainbridge* and the *Halyburton*—and the amphibious assault ship, the USS *Boxer*, set out to the Horn of Africa to rescue the American captain. Add reconnaissance aircraft and the eclectic armada of the pirates, four foreign vessels complete with 54 other hostages from China, Germany, Russia, and several other countries, and a worldwide press corps, and the making of an international incident was complete.

Our first glimpse of Captain Phillips's ordeal and ultimate rescue came from a P-3 Orion surveillance aircraft hovering above the small lifeboat, which was still tethered to the *Bainbridge*. In addition to these live visual images, the FBI hostage negotiation team initiated a regular flow of information: Just as a pebble dropped into a still lake generates ever-widening circles, news stories spread from the orange lifeboat until it reached clear across the globe to Phillips's hometown of Underhill, Vermont, where his wife, Andrea, looked outside her front window and watched as an international press pool grew and widened.

Meanwhile, the pirates were sending out their own message to the world: "We are safe and we are not afraid of the Americans. We will defend ourselves if attacked," one of them told a Reuters reporter via satellite phone. One of the four pirates, Abduhl Wal-i-Musi, was aboard the USS *Bainbridge*, where he received medical care for the injuries he received while fighting with the crew of the *Alabama*. And, like pirates of old, the stakes are high—life and death (i.e., walking the plank)—but so is the treasure. When they are successful, the pirate booty for a single ship can be as much as $2 million—an almost unthinkable amount of money in a desperately poor country like Somalia.[14]

Most countries don't take military action against pirates because they are afraid the crewmen will be harmed. That's why when the United States made that decision, it was such big news: Authorization of the action moved along traditional military channels until it got to the top where U.S. President Barack Obama approved the use of lethal force (i.e., the pirates would be killed) if it was determined that Captain Phillips's life was in danger. The commanding officer on the scene was Vice Admiral William E. Gortney, and he directed Frank Castellano, captain of the USS *Bainbridge*, who determined that Phillips's life was indeed in danger. Castellano directed a team of SEALs to "use the lethal force" in an attempt to save Phillips. Their plan involved a small team of Navy SEALS, who parachuted into the water near the *Halyburton* on Friday afternoon and immediately started a 24/7 surveillance of Phillips and his captors.

They watched and waited day and night using high-powered night vision glasses, constantly monitoring the actions of the now exhausted captain and the close-to-panicked pirates. When the pirates' heads poked above the orange life-boats roof early Sunday morning, April 12, 2009, the SEALs seized the opportunity and quickly fired a volley of shots. The almost delirious Phillips later recalled that he thought the pirates were shooting at each other and it wasn't until he heard an American voice say, "Are you okay?" that he knew he was safe.

The SEALs put Phillips into a nearby inflatable boat and quickly transported him to the waiting USS *Bainbridge*. Instead of exhaustion, fatigue, and the remnants of his harrowing ordeal, we only saw joy and exhalation on Phillips's face when he posed for the now famous picture of him and Navy Commander Frank Castellano before being examined by physicians on the USS *Boxer*. Americans, and indeed the world, breathed a collective sigh of relief as the photo of the smiling captain appeared with headlines of his heroic rescue and ordeal quickly travelling the Internet and appearing on television screens and the covers of magazines.

How Is Our Hunger for Awareness Satisfied?

This is so quintessentially news: The failed hijacking of the *Maersk Alabama* by pirates brandishing AK-47s halfway around the globe in a remote corner of the Indian Ocean is truly news that captures and sustains our attention. Captain Phillips's story had all the "right stuff": conflict, high drama, sophisticated technology, an exotic and remote locale, international intrigue, and the brandishing of the political and military power of the United States and its president.

We hear a radio report, read a news story, see a television clip of the captain and his frightened wife back home in the New Hampshire mountains, and we're hooked. We tune in and stay tuned in for days, months, and even years perhaps. We are curious: we want to know what will happen next. Perhaps we see ourselves in the captain, who bravely faced his own death and survived to tell us what it was like to be him.

Journalism historian Mitchell Stephens calls this innate desire a "hunger for awareness" that when satisfied, helps people feel at home in the world and comfortable that they understand their place in it. Stephens defines news as "new information about a subject of some public interest that is shared with some portion of the public."[15] News outlets offered a great deal to satiate this desire. From the first reports of the hijacking of the ship on April 8, 2009, and the captain's return home to Vermont, to first-anniversary recap stories and the release of Phillips's book, reporting about the Somali pirates dominated news coverage for many months.[16]

Historically, stories about pirates, crime, and dramatic rescues (such as the young child trapped in a drain pipe for days, and many others like it) attract attention and satisfy both innate curiosity and an organic need to know. John Campbell was one of the country's first journalists, and his 1704 *News-Letter* contained information about privateering and piracies, the arrival of ships, deaths, sermons, political appointments, storms, Native American depredations, counterfeiting, fires, accidents, court actions, maritime news, and a few advertisements. Modern reporting has evolved from early journalists like Campbell, who attempted to put the events of the day in some kind of order so the world was simply easier to understand. They acted as "village explainers," members of the community who observed, recorded, and shared everyday events and put them in context. There is an innate and natural tendency to reach out and find out how the world works and what it's like for others to be in the world. That is why news has traditionally been framed as a story, which according to Stephens, follows a narrative arch, which like its musical cousin "plays intentionally on connections to human experience, just as seven-tone music counts on the states of tension, unrest, and resolution it excites in listeners."[17]

Journalists Are "Watchdogs" Guarding Against Abuses of Power

News is also, quite simply, new information that individually and collectively adds to public knowledge and helps people understand their world. Almost instinctively we pay close attention to what is new, what has changed, and what is emerging. Typically, news originates in three different places: naturally occurring events, such as disasters and accidents; planned activities, such as UN General Assembly meetings; and reporter's enterprise, watchdog stories that uncover corruption and abuse of power. Reporter's enterprise is often assumed to be large, ongoing investigations at established organizations, such as the uncovering of the Watergate scandal by Carl Woodward and Mark Bernstein of the *Washington Post* or the more recent *New York Times* exposé on the use of retired Pentagon officials as experts on Sunday morning television talk shows. But dedicated reporters routinely demonstrate enterprise, no matter what the scale. While working as an Eastern Michigan University journalism intern for the *Ann Arbor Chronicle* in March of 2010, Helen Nevius Adamopoulos filed a Freedom of Information Act request (commonly known as FOIAs in newsrooms) and discovered that the board of trustees of Washtenaw Community College had spent more than $10,000 of public funds in one weekend, including a $4,000 dinner, as they luxuriously contemplated a multimillion-dollar facilities expansion.

The responses to Nevius Adamopoulos's investigative reporting posted to the *Chronicle's* website were as significant as the reporting itself. On March 21, Lou Glorie wrote, "Thanks *Chronicle* for a peek behind the curtain. The WCC board needed a full day of closed sessions to discuss a real estate deal? They were wise to take this meeting out of the county where none but the most intrepid would follow. Distractions must be evaded at any cost—close to $10K at latest count. Citizens demanding daylight are indeed a distraction." Vivienne Armentrout also thanked the intrepid Adamopoulos for "bringing us the information in the story. It was very revealing. Unfortunately, if most official discussions are conducted in private, we will not be able to evaluate moves taken by public bodies that will 'impact others.'"

This is journalism at its best: reporters acting as our eyes and ears carefully crafting and sharing stories from across the globe and across the street, and readers engaging with and responding to those stories. It is so critical to have both, not just the reporting but also the active engagement with reporting and the reporters who produced the story. Digital technologies have certainly facilitated this process. Skillful reporters act as surrogates patiently but persistently attending long, often boring, meetings and intelligently questioning the wisdom, actions, and especially the words of elected officials. This is journalism's highest calling—acting as the nation's fourth estate watching for abuses of power. From budding young journalists on a local scale, such as Helen Nevius Adamopoulos, to seasoned veterans such as the *New York Times's* David Barstow, who won a 2009 Pulitzer Prize for his reporting about the use of retired Pentagon generals as experts, reporters are trained and often instinctively recognize abuses of power. It is important that they do their job well and also that we pay close attention and support their endeavors because over time, news stories become our collective histories and our cultural memory.

Reporters Act as Our Eyes and Ears

Historically and by necessity, the public has relied almost exclusively on the discretion, experience, and professional integrity of editors and reporters to determine which of the myriad of events that occur each moment of every day in our lives and around the world are significant enough to become news. As our representatives in foreign and often inaccessible locales, reporters have traditionally decided what to show the public and filtered what they see through their professional lenses. Many of them, such as Edward R. Murrow, Bill Moyers, and Lara Logan—to name just a few—have become heroes and heroines. We invited them into our living rooms and they traveled with us in our cars. For several generations of Americans, they represent what reporters should do and what journalism should be.

The standards and qualities that reporters most commonly use to qualify events as news—timeliness, proximity, prominence, oddity or singularity, conflict, and controversy—come from the mid-1800s, a time of revolutionary change in American journalism when the ideal of objectivity emerged. Before the 1830s, newspapers were partisan, expensive, and exclusive: They were commonly sold by subscription to the wealthy. However, the creation of the Associated Press in 1846, the nation's first wire service, and the advent of the penny press, when cheap daily newspapers were sold on the streets to common folk, forever changed newspapers and the concept of news. News developed a more objective orientation. Stories were told in a factual, direct way so they would appeal to and meet the needs of many different audiences simultaneously.[18]

Storytelling Is Very Different Now

Looking at stories such as the 2009 and 2010 reporting of the attempted hijacking of Captain Phillips's ship, the *Maersk Alabama*, through the historical and theoretical lens of journalism scholars, such as Michael Schudson, Bill Kovach, Tom Rosenstiel, and others, it becomes very apparent that some things have changed. For example, the technology used to produce news stories and deliver them almost immediately to global audiences and certainly the sophistication and growth of potential audiences is dramatically different from earlier times. Yet some things have stayed the same. Pirates are still stealing ships and news organizations continue to construct and frame stories for mass consumption. Many news stories, such as Captain Phillips's adventure, start with a strong and genuine news quality, such as timeliness or conflict. However, beyond the original reporting of the news, in this case, the hijacking, capture, and rescue of Phillips, the story often "de-evolves" as its components are dramatized, emphasized, and reenacted to first attract, then entertain and ultimately sustain the attention of large numbers of people.

"News media are geared to own a story. They shape it, package it, and sell it," explain new media visionaries Shayne Bowman and Chris Willis.[19] And if there is any doubt about the packaging of the Captain Phillips story, it is apparent that the release of his book, *A Captain's Duty: Somali Pirates, Navy SEALs, and Dangerous Days at Sea* (Hyperion, 2010), and the publicity surrounding it was timed to precisely fall on the first anniversary of his rescue. The *Los Angeles Times* reported that Phillips negotiated a movie deal with Columbia Pictures almost immediately after his release. Focus quickly shifted from Phillips's dramatic rescue and safe return to, "Who will play the lead role?" in the movie version of the story. (In the spring of 2011, CNN reported that Tom Hanks will be starring in the Columbia Pictures's version of Phillips story and book.[20])

From the very beginning, Phillips controlled access to his version of the story. NBC featured an exclusive interview with Phillips in April of 2010 and used the interview as the basis of a high-profile retrospective prime-time feature, which focused on Phillips's remembrance of the event, President Obama's decision to use the Navy SEALs, and computer simulations of the actual event. The interview was compelling and dramatic and appealed to the voyeur in all of us. We wanted the details about what it was really like to be him during the longest four days of his life: what was he thinking, what was he feeling, and how did he make the decisions he did? NBC producers understood these almost primal longings and began to satiate our "hunger for awareness" with promises to reveal the "hour-by-hour details of the captain's battle of wits with Somali pirates." NBC's Matt Lauer assured us that the captain would tell "the harrowing true story of those four grueling days on a tiny lifeboat, a tale of terror, despair, and ultimately quiet courage."

We would also hear from President Barack Obama. NBC described the Phillips rescue as "one of the first foreign crises of his young Presidency."

Phillips consistently delivered as promised. "I was afraid for the entire time," he revealed. Studio shots of Phillips and Lauer were juxtaposed with sophisticated computer animations and a small number of U.S. Navy video and helicopter surveillance photos to help the audience better understand the sequence of events the captain described and the logistics of the hostage scene and rescue.[21]

Lauer is the quintessential guide and narrator. As the *Dateline NBC* version of the story unfolded, he interrogated Phillips and created a faux Socratic dialogue asking the cooperative captain sequential but leading questions that both move the story forward and clarify Phillips's state of mind and his decision-making processes during the ordeal. This is demonstrated very clearly in perhaps the most dramatic episode of the saga—Phillips failed escape. Here's how Lauer and the NBC *Dateline* team told that part of the story:

> LAUER: (Voiceover) He saw that one of the pirates was relieving himself off the side while the others were sleeping.
>
> (Ocean at night) [visual]
>
> LAUER: You've got two sleeping in one part of the lifeboat. You've got the other guy dozing off in the conning tower, and this guy's put his gun down.
>
> Capt. PHILLIPS: And then I just said, "This is it." I pushed him. Had to push him twice. He went screaming into the water, which woke the other pirates. And for a second I was like frozen with that weapon there. But I didn't know how to use it. I put it away, and I dove in.

Remember now that we do know how the story will turn out! The rescued captain sits before us after all, and the retelling of his story is structured and staged to create maximum impact. If we want to know what it was like to be Phillips, we really want to know what it's like to be the president, who also explained what helped him make the difficult decision to use deadly force to rescue an American hostage. That is why as the dramatic arch of the story rises and we await further details and descriptions, the NBC *Dateline* team brings in the most credible source of all—the President of the United States Barack Obama. Obama's statements are adroitly sprinkled throughout the program as well as its promotional clips. After stating his concern for Captain Phillips and his family, President Obama explains his decision-making process and rationale for supporting the use of deadly force to rescue the cargo ship captain.

> Pres. OBAMA: You know, one of the things you discover as commander-in-chief is that our military is so well-trained that with confidence, I can give them some clear instructions with some clear parameters and have confidence that they will exercise whatever discretion they need effectively.

The President calmly assures the American public that the government will use its military power to ensure the safe rescue of its citizens: "Immediately we went to work trying to coordinate all US power to figure out how we can free him," he said.

"News About News" Misinforms

That specific remark by President Obama was repeated frequently throughout the hour-long show as were key phrases of Phillips's, such as "all hell broke loose," "taking turns hitting me," and "with guns pointed at his head." Numerous pieces of information and images were used multiple times throughout the show and in its network promotion. In ways similar to other "docudramas," the story was not constructed to deliver news and inform audiences but to attract and sustain attention. For example, before each commercial break, we are reminded again of the real and implied danger to Phillips's life during his four-day captivity. The only new insights are about Phillips's thinking at the time—what was it really like to be him and to be afraid for your life. There was little or no new information beyond Phillips's interpretation of events, which he had mostly withheld for a year until his book was released.

By focusing almost exclusively on the unfolding of the story from the captain's internal point of view, *Dateline* excluded and neglected other significant questions about historical context, international commerce, and legal and military procedures and processes that engaged and attentive audiences would naturally bring to this story. For example, why do these cargo companies carry expensive insurance instead of hiring armed and trained security guards? These important contextual issues were directly addressed by none other than satiric comedian and news observer Jon Stewart.[22]

Briefly, Stewart challenged *The Daily Show's* "senior maritime analyst" Briton John Oliver on his ambivalence regarding the Navy SEALs' rescue of Captain Phillips. Oliver's satiric response succinctly summed up some of the more complex and critical issues surrounding the dramatic rescue: "I, and indeed, much of the world, are naturally uncomfortable with ostentatious projections of American power given the morally questionable ends to which such power has been wielded in the past, from the Philippines, to Central America, all the way through to modern-day Iraq," said Oliver. Using the guise of comedy, Stewart and Oliver directly addressed a vitally important part of the news story about Phillips's rescue and directly confront the role and perception of American military aggression around the world.

Again, in their witty and satirical style, Stewart and Oliver also poked fun, but in so doing drew attention to the significant financial investment news organizations make in creating computer simulations, as John Oliver unveiled his sophisticated high-tech computer simulation—a laptop acting as a ship floating in a bathtub with plastic boats and action figures. "Are you sure you want to use a really expensive computer?" Oliver asked Stewart, who responded, "Yes, let's do it. It would be really cool."

While the significant historical and political context of the story of Captain Phillips's rescue was missed by the *Dateline* reporting of Matt Lauer and his news team—even after they had a year to do a substantial amount of reporting—it was clearly a part of the newsphere, as is evidenced in the following comment posted to *The Daily Show's* website:

> Jon, I love the show, I'm a big fan. BUT, this pirate thing is a disgrace. Almost all, if not all, of the mainstream media (and you, too!) reported this as if this was some great heroic rescue mission—saving a US Captain from those evil pirates—when the truth is THOSE PIRATES WERE SIMPLY POOR FISHERMAN WHO BECAME PIRATES AFTER YEARS OF FOREIGN COUNTRIES (THE USA AND EUROPE) STEALING THEIR FISH AND DUMPING TOXIC NUCLEAR WASTE OFF THEIR SHORELINES! You are a fisherman, you are poor but making a living, and then foreign countries/companies steal your fish and dump toxic nuclear waste on your shores… now your people are sick and you cannot make enough money or get enough food to survive. So…you become a pirate and attack the a-holes who are doing this to you. In that light, capturing a US Captain who is supporting this whole **** parade sure looks different, right? And then you do this "Pirate Rescue Simulation" bit as if it is a heroic act to shoot three of these poor fishermen in the head to save one US Captain! I would argue it is NEVER heroic to shoot three people in the head, much less poor fisherman just trying to survive the onslaught of capitalist imperialists. Terrible. Please do another bit on this to correct this! Other than that, the show is great. Peace, Peter[23]

As John Oliver and Peter (aka HappyPedro), the blogger who posted this comment on the Comedy Central blog, pointed out, there are many sides to this multifaceted international story that were sorely missed by mainstream outlets, especially the *Dateline* prime-time special. Traditional qualities of news—fairness, balance, comprehensiveness, among others—were missed by *Dateline* in its quest to dramatize the story. Stewart and Oliver identified several important contextual issues important to the Phillips rescue story. Johann Hari, who wrote "You Are Being Lied to About the Pirates" for the *Bay View*, a web-based news outlet in San Francisco, also provided background and historical perspective. "Pirates have never been quite who we think they are," Hari reminds his readers as he retells the story of the marauding Somalis:

> In 1991, the government of Somalia—in the Horn of Africa—collapsed. Its 9 million people have been teetering on starvation ever since—and many of the ugliest forces in the Western world have seen this as a great opportunity to steal the country's food supply and dump our nuclear waste in their seas.…
>
> At the same time, other European ships have been looting Somalia's seas of their greatest resource: seafood. We have destroyed our own fish stocks by over-exploita-

tion—and now we have moved on to theirs. More than $300 million worth of tuna, shrimp, lobster and other sea life is being stolen every year by vast trawlers illegally sailing into Somalia's unprotected seas. This is the context in which the men being called "pirates" have emerged. Everyone agrees they were ordinary Somalian fishermen who at first took speedboats to try to dissuade the dumpers and trawlers, or at least wage a "tax" on them. They call themselves the Volunteer Coast Guard of Somalia—and it's not hard to see why.[24]

In contrast to the *Dateline NBC* story that focused almost exclusively on the rescued captain's point of view and a retelling of the sequence of events surrounding the capture and rescue of the captain, Hari's writing and reporting in the San Francisco *Bay View* powerfully describe another aspect of this story. Hari's tone is highly subjective, but his reporting is well-sourced and he relies on credible experts, such as Ahmedou Ould-Abdallah, the UN envoy to Somalia as well as conversations with the Somalis.

In sharp contrast to Lauer's celebratory tone and Stewart's humor, he attempts to explain why the Somali people have resorted to piracy. His insights and reporting go much deeper than Lauer's because he provides readers with the story's important historical, political, and cultural background.

There is a power and relevance to his prose here: in fact, more than 300 readers were so engaged by the story that they posted comments to the *Bay View*'s blog. The variety and often vehemence of these posts is enlightening and points to an important principle about the newsphere: Agreement is not a necessary condition for the truth to exist. It often feels that the receivers of news and information first process what they read or hear through their own personal filters and then (often quickly) decide whether the information is true. They then proceed to share their "newly arrived, personal, and original" truth with others. This is a perfectly acceptable practice, but it is not news dissemination. There is an objective truth and an objective reality independent of our perception and our own personal reality, and quality journalism ideally represents this truth. As these examples and this analysis shows, discovering and uncovering this kind of quality journalism that represents the truth is especially challenging now.

Journalism Is a Dynamic Search for Truth

In their *The Elements of Journalism: What News People Should Know and the Public Should Expect*, Project for Excellence in Journalism's senior counselor Bill Kovach and director Tom Rosenstiel devote an entire chapter to truth—"Truth: The First and Most Confusing Principle" of the ten elements they describe. They outline the challenges to the journalistic discovery (indeed all discovery) of the truth, such as sophisticated tools for information management, the glorification of perception as reality, and confusion about objectivity. They describe journalistic truth as more than mere accuracy. Instead, it is a "sorting out process that takes place between the initial story and the interaction among the public, newsmakers, and journalists."[25] Kovach and Rosenstiel acknowledge that journalism "exists in a social context" and so the result of that process—news—evolves according to the dynamic interchange of the public and the product of journalism—news.

It follows then that while a story can be fair and accurate, real truth evolves over time in large part based on the demand placed on it by the people who consume and use it: It is the value we bring to journalistic fare—our own individual and collective demands for relevance, and yes, truth—that will ultimately result in the quality we deserve. That is because "journalistic truth is a process," and a "continuing journey toward understanding—that begins with the first story and builds over time."[26] In the past, that journey was mostly taken by reporters, editors, publishers, and producers. Now that a whole host of ever-evolving two-way digital tools exist, we are no longer spectators but powerful participants in this process. Reaping the benefits of journalistic truth means now joining in this critical process.

Much of what may feel like real news in the newsphere today is simply someone trying to persuade someone else (or multitudes) to accept their belief system and interpretation of a certain set of facts or circumstances or a spurious debating of facts. The existence of voluminous interpretations of reality and the singularity of individual reactions to news is beautifully illustrated by the responses to Hari's story about the Somali pirates. More than 300 readers were inspired to post on the San Francisco's *Bay View* blog. The variety, vehemence, and violence of these remarks is both startling and informative. Here is a brief sample of comments in support of Hari's argument as well as the angry responses that live a long life on the Internet when sites do not moderate comments.

The wide range of responses illustrates the diversity of newsreaders in the newsphere. Many of the comments to Hari's article were reasoned and appropriate and those commenting used their own names, such as Robert Rasmussen. Rasmussen posted the following (excerpted here):

> I've been studying this issue as a security studies student for the last two years. I like what a Commander of the Royal Navy once said.... "In order to eliminate piracy, one has to eliminate the problems of Somalia."
>
> What this article fails to mention is that Somali piracy started out as an angry reaction to illegal fishing and chemical dumping, but eventually evolved into a business-profit making venture that is in many cases controlled by warlords, who in some cases conscript local fishermen to become pirates. This is why in the last three years attacks have skyrocketed to include large shipping vessels and cruise liners. Not to mention that much untraceable capital in an environment with that much anarchy breeds the perfect region for a criminal-terrorist nexus (interchange point) to occur...although there is no proof one way or another.

Not all of the comments were as respectful in tone and content as Rasmussen's. Here's what RU Kidding and Jazzykat, who participated in heated exchanges with other bloggers posted:

> "Leave it to the kooks in San Francisco to come up with a Blame America First slant on Thuggery and Thievery." —Posted by RU Kidding

> "They need to send more than a few black hawk helicopters—maybe some nuke tipped cruise missels would do! I'm of the old school where we didn't take crap off of anybody (where strength creates peaceful neighbors—it certainly worked with Japan after we nuked them) …" —Posted by Jazzykat

Moderation most often solves the problem of inappropriate commenting, such as these and unfortunately many others. Unlike face-to-face communication, bloggers too often used the anonymity provided them to spew hateful language and ideas and spread them to others. Yet when it is used responsibly, as Robert Rasmussen has done here, posting informed comments in response to a story supports the goal of quality news in the newsphere. Responsible blogging supports engagement—what most journalists covet. In his insightful *Watching the Watchdog: Bloggers as the Fifth Estate*, Stephen Cooper outlines the additive qualities of blogging:

> Bloggers do have the power to identify factual inaccuracies in mainstream reporting, second-guess the news judgment of mainstream editors, argue for different interpretation of facts than those offered in mainstream stories, or draw attention to stories they feel have been insufficiently covered.[27]

Blogging is a way of following a story, and as we will see in the upcoming chapters on building and filtering news networks, monitoring the blogosphere to find the most useful blogs is a key strategy for conscious consumption.

Journalist as Blogger: A New Role for Professional Journalists

News has both evolved and, as examined in this chapter, de-evolved despite sophisticated advances in new digital tools. However, when used properly, the open exchange of information and the prompting to continue often difficult searches for the truth offer unprecedented opportunities now. In the hands of dedicated and skillful professional journalists, such as NBC's Jim Miklaszewski, blogging is a powerful tool for reporters. Miklaszewski, NBC's 25-year veteran Pentagon reporter, posted that "the military sought and President Obama immediately granted the request and authority for SEAL TEAM SIX to use 'lethal force' if it was determined Captain Phillips life [*sic*] was in danger."[28] He explained that presidential authority is required and was granted to move two SEAL teams—a nearby team in Kenya and ultimately the more advanced SEAL Team Six from Virginia—to the *Bainbridge* in the Indian Ocean.

Miklaszewski's reporting here and also Joe Miller's investigation on FactCheck. org[29] were two responses to an e-mail circulating on the Internet under the signature of Rear Admiral Lou Sarosdy that the president was delaying the decision to use the SEALs, who ultimately rescued Phillips.[30] Several reporters tried unsuccessfully to trace the origins of the e-mail, which appeared to have been written by a SEAL because of the expert knowledge it revealed. The e-mail reads, in part:

> BHO (President) wouldn't authorize the DEVGRU/NSWC SEAL teams to the scene for 36 hours going against OSC (on scene commander) recommendation. BHO immediately claims credit for his "daring and decisive" behavior.

Is tracking down rumors such as these and reporting on their validity news? Yes and no. It has always been the diligent reporter's job to track down leads and see where they go, but doing that is very different now. And as is evident here, the newsphere and the blogosphere can be used for good or for ill. Professionals as well as non-professional reporters and the population of the blogosphere need to work together to pursue the truth, but the verbiage created around that verification— what I call "news about news"—is not actually news. What is valuable is the original reporting that supports the primary story—not the reporting that repeats the challenges to the story. And if a story is picked up in the middle, say, only from the discussion of President Obama's decision to use force, then it is very difficult, perhaps impossible, to distinguish between original reporting and "news about news."

If you enter a story at the middle or the end, it takes a conscious effort to separate the primary story from the challenges to the story that appear to present new reporting of information. The kind of news, then, that Miklaszewski and Miller are creating and reporting is organically different from Lauer's *Dateline* interview with Captain Phillips. The key difference in these "styles" of news is the amount and quality of new information—the access to and delivery of real reporting of new information to the public via the newsphere.

We move next to better understanding the benefits of courageously uncovering and sharing the truth and the dire consequences of passive acceptance of disinformation and lies.

ENDNOTES

1 (2001). p. 16.

2 Carey, J. (1993). p. 1.

3 PEJ. (2007).

4 See above. PEJ. (2007).

5 http://www.opednews.com/articles/Photo-Christians-Protecti-by-the-web-110203-446. html. Retrieved February 4, 2011.

6 Schudson. p. 14.

7 Carey, J. (1996). pp. 1–2.

8 Carey, J. (1989). p. 17.

9 See previous note. Carey, J. (1989). p. 21.

10 Schudson, M. (2003). p. 54: "The power of the media lies not only (and not even primarily) in its power to declare things to be true, but in its power to provide the forms in which the declarations appear. News in a newspaper or on television has a relationship to the 'real world,' not only in content but also in form, that is, in the way the world is incorporated into unquestioned and unnoticed conventions of narration, and then transfigured, no longer a subject for discussion but a premise of any conversation at all."

11 Robert Cathcart (1993) builds on the historical understanding of media as channels—"a medium is not only a channel or channels of communication"—and extends the definition to include the context and reciprocal quality of media—"but it is also a learned, shared, and arbitrary system of symbols" (p. 292). His definition is useful here because it also requires us to put form in the foreground as the key to "disentangle the content of modern media from their technical forms" (pp. 304–305). It is exceedingly challenging to separate the medium from the message and the message from its cultural ancestry. Rhetorician Kenneth Burke claims it is impossible to separate form and content. However, Cathcart turns to Burke for the answer to this difficult dilemma: he claims that we now need a Burkeian philosophy of media form to complement his philosophy of literary form (p. 304). It is an insightful point because it challenges all communication scholars and particularly rhetoricians and media theorists to first understand that there are forms "peculiar to each medium," and second to be able to identify those medium-specific forms (Gumpert, G., & Cathcart, R., 1985, pp. 28–29). As these steps are taken, we will ideally evolve a fuller understanding of precisely how news is rhetorically integrated into the medium that supports and delivers it.

12 The complete report and accompanying graphics and details are available at http://www.journalism.org/analysis_report/how_news_happens. Retrieved May 10, 2011.

13 Available at http://www.good.is/post/transparency-the-biggest-news-stories-of-the-year. Retrieved May 10, 2011.

14 Available at http://query.nytimes.com/gst/fullpage.html?res=9500E6DF133CF933A25757 C0A96F9C8B63&pagewanted=2. Retrieved April 29, 2011. On April 10, 2009, the *New York Times* reported that in 2008 alone, insurance and shipping companies paid more than $100 million to get their vessels back.

15 Stephens, M. p. 4.

16 Good.is and journalism.org graphically illustrated the top news stories of 2009.

17 Stephens, M. p. 11.

18 Michael Schudson's 1978 *Discovering the News: A Social History of American Newspapers*, which is based on his doctoral dissertation, historically deconstructs objectivity as a journalistic ideal and standard.

19 Bowman, S., & Willis, C. *We Media: How Audiences Are Shaping the Future of News and Information.*

20 Available at http://www.hollywoodreporter.com/news/tom-hanks-play-capt-richard-167846. Retrieved May 10, 2011.

21 *The New York Times* reported on a now familiar concept, "Maybe Journalism," when Japanese news organizations re-created Tiger Woods crashing his SUV and criticized news reporting "presented as a video game." However NBC's computer simulations of Phillips hostage-taking and rescue are very similar and qualify also as "depictions of events no journalists witnessed." See http://www.nytimes.com/2009/12/06/business/media/06animate.html?scp=4&sq=noam%20cohen&st=cse. Retrieved May 10, 2011.

22 "The Buc Stops Here—Maritime Rescue Simulation" was one of the program's most popular episodes. In July of 2010, episode #14040 had 54,422 views.

23 Please see http://www.thedailyshow.com/watch/tue-april-14-2009/pirate-rescue-simulation

24 http://sfbayview.com/2009/04/.

25 Kovach, B., & Rosenstiel, T. *The Elements of Journalism: What Newspeople Should Know and the Public Should Expect.* p. 41.

26 See Kovach, B., & Rosenstiel, T. p. 43.

27 Cooper, S. p. 14.

28 Please see Miklaszewski, J. (2009, April 21). SEAL email criticizing Obama is bogus [Web log post]. Retrieved from http://firstread.msnbc.msn.com/_nv/more/section/archive?year=2009&month=4&ct=a&pc=25&sp=100#April 2009 archive_nav.

29 Please see http://www.factcheck.org/2009/04/here-there-be-pirates/.

30 Please see "Navy Seals Rescue Captain Phillips from Somali Pirates," available at http://www.snopes.com/politics/obama/pirates.asp.

· 3 ·

CHOOSING THE COMFORTABLE LIE OR THE UNCOMFORTABLE TRUTH

"False and contrary to nature is the egocentric ideal of a future reserved for those who have known egotistically how to reach the extremes of 'everyone for himself.' No element can move or grow unless with and by means of all the others as well as itself."
Teilhard de Chardin, The Human Phenomenon

I was in the Russell Sage Laboratory, a beautiful, old, ivy-covered building on the Rensselaer campus in Troy, New York, teaching "Writing to the World Wide Web," a popular writing course, when Colin Powell gave his testimony to the United Nations in March of 2003 about Iraq possessing weapons of mass destruction. The class was discussing "ethos," one of Aristotle's three modes of persuasion, and the use of then-Secretary of State Powell's formidable character and accomplishments as the primary mode of persuasion to convince the UN delegates that the United States was indeed justified in waging war against Iraq.

Rensselaer's bright and industrious undergraduates easily identified the rhetorical flaws in Powell's case—his evidence was faulty. Despite the diminishment of quality from the transmission of the photos onto the classroom's desktop computer screens and their further degradation upon projection, it was very clear that the aerial photos Powell was using as evidence were limited: They did not really show anything except the tops of buildings. We enjoyed a lively discussion as I again witnessed the joys of learning from my students, who concluded that it was the rhetorical power of Powell's character and intelligence that convinced his audience that day.

Several years have passed and so it is now easy to look back upon that day with close to 20/20 vision: There were no weapons of mass destruction and Powell had used his influence to convince the world of the rightness of American aggression in the Middle East. And it may seem at first glance that this is a simple story of the mass media getting it wrong again, or the government needing to take strong action to protect American interests (as some might believe), or simply a failure of U.S. intelligence, as some claimed. But, as is most often the case, the story is much more complex. While it may appear that most American journalists got the story wrong—they did not act as our watchdogs against abuse of governmental power—several of them nailed it. *And perhaps we need to look at ourselves and why many of us missed this important message.*

One of the news outlets that got the story right is *Vanity Fair*, an eclectic mix of fashion, photography, celebrity gossip, and hard-hitting investigative journalism—the kind that many experts say is lost. You might say that PBS's Charlie Rose also got the story right because he interviewed the team of reporters from the U.S. and other countries that the magazine had assembled to investigate the story in advance of its publication in the fall of 2004. Rose, in his low-keyed but powerful style, quickly got to the reporting behind the story and explored how *Vanity Fair* hired expert investigators and the best reporters, who spent much of their time in London because sources here in the U.S. were inaccessible.

Titled "The Path to War: Special Report: The Rush to Invade Iraq—The Ultimate Inside Account," the article appeared on the cover of the May 2004 magazine along with a picture of Jackie Kennedy that accompanied a "Private Camelot" article. More than six months before the 2004 national election, *Vanity Fair* published this hard-hitting, factual accounting of Powell's discovery that there were indeed no weapons of mass destruction and the premise for the U.S. invasion was flawed. Not only was the timing of this article impeccable, but its authors, Bryan Burrough, Evgenia Peretz, David Rose, and David Wise, got the story right when many, many very smart people got the story wrong, and, perhaps most tragically, unwittingly perpetuated a lie. And in fitting with *VF's* style, the story was also impeccably accessorized by Annie Leibovitz's photographs of the Bush Cabinet.[1]

Vanity Fair was not the only news outlet to get the story right. Reporters at the McClatchy news service also accurately reported the story. So if evidence existed that there were no weapons of mass destruction, why didn't this information travel through the newsphere more efficiently and effectively so the American public might have used this knowledge to make more informed choices during the 2004 Presidential election?[2] The question and attempts to answer it are complex; this chapter explores what constitutes quality journalism, and more importantly, whose job it is to find that journalism and use it well.

One of the clearest, most passionate critics of the current news structure in America—a witness who worked on different sides of the news desk and actively and passionately pursued meaningful change—is Bill Moyers. In April of 2007, Moyers left a comfortable retirement to fund-raise in support of his PBS television show, *Bill Moyers Journal*. The show aired weekly until April of 2010. In its debut episode, "Buying the War," Moyers told the story of the media industry's role in helping the Bush administration convince the American people that the Iraq War was necessary to secure our national defense. To appropriately bookend his legacy to American journalism and celebrate his retirement from the *Journal*, Moyers carefully selected a guest for his April 30, 2010, final episode. In addition to a report on the "Iowa Citizens for Community Improvement" and an interview with community organizer Jim Hightower, Moyers concluded the show by talking to Barry Lopez, an Oregon novelist and writer whose work captures "one man's effort to go out into the world, to discover what is beyond and within us."[3]

Moyers shared his reasons for selecting Lopez from among the hundreds of other men and women who inspired him. "His curiosity about the world and pursuit of it have set the gold standard for all of us who have dedicated our lives' work to explaining things we don't understand," Moyers told his audience that April night. In many ways, Lopez is a kindred spirit to Teilhard de Chardin. Both are Jesuit-educated and both see the Earth as a living organism: "The wind is alive. It has a soul. It's part of the moral universe," Lopez told Moyers. And like Teilhard, Lopez believes in action, acceptance, and our organic interconnectedness: "The play is not directable. You must participate in the play. You must get out of the director's chair telling everybody what to do and how to behave and who can be on stage. You must put all that aside and step onto the stage with other men and women and say, 'We're all in this together. And we need to find an arrangement… in order to take care of each other.' But we can't exclude. We can't make nature the banished relative, no part of the human family."[4]

This is the cosmic sense that Teilhard refers to—a connection to le Tout (the All). Many news stories have lost this essence—this ability to inform, elevate, and unite. Often, instead of bridging differences and building a connection to others, news reporting supports and confirms disconnection—making people feel separate from the world and not a part of it. And worse, constant reminders and the "up close and personal" exposure to human atrocities, failures, and pure evil that often constitute news coverage can induce and prolong despair. In ways similar to Teilhard, Lopez acknowledges the inherent dualities in the human condition that require us to "lean into the light," and to maintain a sense of hope: "How can we maintain our sense of hope when to go deep into the news is to encounter the kind of terror that can traumatize a person for the rest of their life? I think hope is a space holder for that word," Lopez told Moyers.

In bringing the life, work, and wisdom of visionaries, such as Barry Lopez, to public attention, Moyers has built on his longevity, stature, and unique vantage point to cast a clear and compelling focus on some of the journalism profession's most pressing and complex challenges. In his 1989 series *The Public Mind*, Moyers dramatically illustrates how an individual's innate perceptual bias is compounded and distorted when elevated to the collective of the newsphere. Moyers' work dramatically and eloquently points out the blind spots in the collective mind: "America often refuses to face the painful truth about itself," he said (before we discovered that there really were not weapons of mass destruction in Iraq and the country launched into another war). And in true Moyers style, he moves beyond blame and criticism to get to the root of the real issue: denial.[5]

In "*The Public Mind*" video series, Moyers builds on the progressive and pioneering work of behavioral scientist Paul Ekman, an authority on deception and one of the 100 most eminent psychologists of the twentieth century. Ekman is a retired professor from the University of Southern California, San Francisco, and is also known for developing the character Cal Lightman, portrayed by the actor Tim Roth on the Fox television series *Lie to Me*. In his interview with Ekman, Moyers reminds us that lies would not exist and thrive in the newsphere if people did not believe them. Our belief keeps lies alive: "Every lie in the public mind needs a willing listener…needs a listener or listeners who want to believe that it's true," Ekman told Moyers.

The collective mind has its own blind spots and works in a parallel way to the individual mind: It is the sum total of the things not looked at because they are too anxiety-provoking and too painful, according to Ekman. He calls these primitive deceptions "vital lies," that "we need in order to live and keep ourselves from being overcome by anxiety." The real danger, especially when elevated to the collective level, is that "you don't notice that you don't notice"—lies are so deep and so primitive that their existence is no longer recognized: They look and feel like the truth.

Moyers poignantly concludes his *Public Mind* series with a powerful call-to-action for journalists and "journalist wannabes" alike: "We seem to prefer the comfortable lie to the uncomfortable truth. Have the lies become so many that the public mind has become saturated and numb by them, or have we actually become addicted to deception? We punish those who point our reality while rewarding those who provide us with the comfort of illusion. Reality is fearsome, but as we learned from this series, experience tells us, more fearsome yet is evading it."[6]

Flaws in the Hearts and Minds of Journalists and Newsrooms

One of the most celebrated and astute depictions of the realities of writing and reporting news is Robert Darnton's essay "Writing News and Telling Stories."[7] Darnton, an emeritus history professor at Princeton University, studied at Harvard and Oxford, where he earned his doctoral degree. He worked at *The New York Times* from 1965 to 1966, and wrote "Writing News and Telling Stories" as a first-person perspective on the organic bias that exists not only in newsrooms but in the minds and hearts of journalists. "We really wrote for one another," wrote Darnton.[8] He explains in detail what actually happens in the mind of a journalist as he or she composes a story:

> Almost every article develops around a core conception of what constitutes "the story," which may emerge from the reporter's contracts with allies in the city room as well as from his dialogue with the editors. Just as messages pass through a "two step" or multi-step communication process on the receiving end, they pass through several stages in their formation. If the communicator is a city reporter, he filters his ideas through reference groups and roles set in the city room before turning them loose on "the public."[9]

Darnton's article is considered canonical and it often appears on the reading list of undergraduate journalism and communication courses. It was also referenced in Elizabeth Kolbert's "Tumult in the Newsroom," which appeared in the June 30, 2003, issue of *The New Yorker* magazine. In her article, Kolbert likens Darnton's "Writing News and Telling Stories," to an "in-house Masters and Johnson," the famous guide to sexual behavior, because it is considered basic instruction for fledgling reporters at *The New York Times*. Kolbert's *New Yorker* piece recalls one of the darker moments in the history of *The New York Times* when the organization was painfully embarrassed by accusations that a former reporter, Jayson Blair, repeatedly invented sources and stories that the *Times* published.

Kolbert's depiction of the *Times* is less than flattering: She characterized it as a "rigidly hierarchical, almost neurotically status-conscious place" driven by the "competitive metabolism" formula promulgated by former editor Howell Raines, who resigned following the Jayson Blair scandal. Both Darnton and Kolbert point to the need for a more evolved form of journalistic transparency as a partial solution to both the individual and systemic bias—and also human fallibility—inherent in newsrooms and in the social and cultural practices of news generation. As we have touched on earlier and will look at closely, many of the challenges confronting news construction, distribution, and use stem from misuse of power.

John Fiske, professor of communication arts at the University of Wisconsin–Madison, is one of the leading experts on reception theory and the complex interchanges that result when power enters the equation of meaning making and knowledge circulation. In his *Reading the Popular*, Fiske, who is known for coining the term "semiotic democracy," looks at the forces behind news design, dissemination, and consumption, and he makes a strong case for the open networks explored in upcoming chapters. "Knowledge is power," says Fiske, and he believes that the circulation of knowledge is part of the social distribution of power: "Discursive power involves a struggle both to construct a (sense of) reality and to circulate that reality as widely and smoothly as possible throughout society."

In the news environment, this struggle occurs on two levels: The first is the determination of what is "real" and the establishment of precisely what we can truly know. The arbitrariness and inadequacy of this "must be disguised as far as possible," according to Fiske.[10] The second level of this powerful discursive struggle involves persuading and convincing others, "whose interests may not necessarily be served by accepting" the truth of a particular construction of reality. So instead of focusing attention on objectivity, accuracy, or bias and how well events are represented in the newsphere, Fiske suggests focusing instead on "different strategies of representation, between legitimated and repressed senses of the real" to better understand and truly appreciate what and who determines the reality of the news.[11]

To promote more active engagement of audiences with the news and information they encounter, Fiske recommends that the news producer "discard its role of privileged information-giver, with its clear distinction between the one who knows (the author) and those who do not know (the audience), for that gives it the place and the tone of the author-god and discourages popular productivity."[12] Instead of the current traditional role, he recommends that news producers encourage active and open-ended meaning making and help audiences to integrate reported events into their lives: "When the events of the news become woven into the fabric of everyday life they are made to matter," says Fiske. "Instead of saying, "This is what happened today," news reports should say, "These are the events we are in the middle of.""[13]

As may be obvious at this point, Fiske's framework directly challenges the journalistic myth of objectivity: "Objectivity is authority in disguise," according to Fiske, who observes that "facts can always support particular points of view and their 'objectivity' can exist only as part of the play of power."[14] Fact-checking websites and organizations, such as the Annenberg Foundation's FactCheck.org, is an important step to build news literacy in this country and a useful tool to counteract negative political ads in particular. However, fact-checking alone does not address the agenda setting, framing, and story selection that come first.

Holding journalism and journalists to standards of objectivity (and their kindred spirits, fairness and balance) implies a "naïve empiricism"—the world can be known and named for us by reporters.[15] When claims of objectivity are made, facts cannot be challenged and the audience becomes a passive recipient of information. In his *Just the Facts: How "Objectivity" Came to Define American Journalism*, David Mindich echoes and elaborates on the observation of other journalism scholars, including Michael Schudson, that objectivity breeds passivity: "It is no less than remarkable that years after consciousness was complicated by Freud, perspective was challenged by Picasso, writing was deconstructed by Derrida, and 'objectivity' was abandoned by practically everyone outside the newsroom, 'objectivity' is still the style of journalism that our newspaper articles and broadcast reports are written in, or against."[16]

In addition to encouraging conformity and disengagement in audiences, using objectivity as a measure of journalistic quality also disempowers journalists because it casts them in the untenable role of passive observer, which is not a fair representation of what journalists do, according to Mindich. "As more and more information is printed, broadcast, faxed, phoned, and sent over the Internet, being a passive mirror is less and less relevant. With so much news and so many storytellers, there must be something more than passivity for responsible journalists to offer," says Mindich. And, as we will see in upcoming chapters, the art of finding good journalism means developing the finely tuned filters that good journalists use to remediate between "out there and in here." What experienced journalists need to do now is to explain the art and science of filter-making—to explain to their publics how they interpret reality and how they have usefully interpreted reality for their audiences in the past. That means, of course, finally abandoning the myth of objectivity and creating new journalistic standards for a post-objective world.[17]

How Public Opinion Is Manipulated in a Democracy

Media ecologists and historians agree that media are mutually constitutive: They shape and are shaped by society. Given this mutually constitutive relationship, it would appear that the form of the medium, then, would logically follow its function (and vice versa). Therefore, if the American media system was created to support the operation of its democratic government, then the country should have a communication system that supports its ideals. However, it appears that the American system of news and information delivery has evolved in the wrong way.

Princeton historian Paul Starr looks carefully at this dilemma in his ongoing study of media in America, *The Creation of the Media*. Here he observes, "If the origins of modern communication had been, in critical respects, liberal and democratic, how then had the media developed along lines that were so deeply in tension with those ideas?"[18] Starr goes on to define the inherent paradox of the American media functioning in its democracy: "This was the American achievement—and the American dilemma—in communications. In the twentieth century, particularly in the aftermath of World War 1 and other developments that raised concerns about manipulation of public opinion, some critics began to ask how to reconcile democratic ideals with the media's power and limitations."[19]

Throughout the last century, the work of three writers and scholars—Upton Sinclair, Noam Chomsky, and Robert McChesney—have powerfully identified, described, and challenged the systemic challenges of the American media. Sinclair self-published his *Brass Check* in 1919, a muckraking exposé of American journalism that he believed was "the most important and most dangerous book I have ever written."[20]

Next is Noam Chomsky, the MIT linguistics professor, cognitive scientist, philosopher, and social activist who is well known for applying the term "propaganda" to the news and information distributed by the American media. Together with economist Edward Herman he wrote *Manufacturing Consent: The Political Economy of the Mass Media* in 1988. Chomsky and Herman's "propaganda model" views the private media as businesses interested in the sale of a product—readers and audiences—to other businesses (advertisers) rather than providing quality news to the public. Their premise is that an advertising-supported media structure commodifies the creation, production, and delivery of news, which is then "sold" to the public to "manufacture their consent" on important social, political, and economic policies.[21]

The American media landscape is now very different than when first Upton Sinclair and more recently Noam Chomsky voiced their criticism. One of the most profound influences on the newsphere today is the advent of digital technologies and the ever-evolving worldwide computer and information networks that support it. Journalism scholar and media ecologist, James W. Carey, believes "we are witnessing another increase in the scale of social organization based upon electronic communication. We are witnessing the imperial struggle of the early age of print all over again but now with communication systems that transmit messages at the extremes of the laws of physics."[22]

Given the power and partially realized potential of these new communication systems, especially the Internet, the democratic postulate of a free and functioning press creating a public sphere where individuals can interact and debate the issues of the day is at grave risk from the conflicting priorities of commercial markets. Unlike their predecessors, radio and television, which promised the enhanced democratization of the public, the Internet and digital innovation have much more power to manipulate audiences. "In times like these, when revolutionary technologies like the Internet hold extraordinary potential for democratic communication, it is imperative that we prevent the present appropriation of digital communication to suit the needs of business and advertisers first and foremost," warns Robert McChesney, the third in our media critics trilogy.[23] McChesney is a communications professor at the University of Illinois at Urbana–Champaign and cofounder with John Nichols of the Free Press, a national organization working to reform the media through education, advocacy, and organizing. McChesney, Nichols, and a host of other advocates working with their organization, the Free Press, are pressing the Federal Communications Commission—the link between the development, regulation, and ownership of new communication technologies and the public—to keep the Internet free and open.[24]

In his writing, McChesney has assailed American newspapers as "anti-democratic forces." He believes the democracy of the Founding Fathers and the ideals they set forth and guaranteed in the *United States Constitution* are not being practiced in this country. In fact, McChesney qualifies the term democracy, adding the adjectives "anti-neo-liberalism" in an attempt to explain the current functioning of the American political system. "If we value democracy, it is imperative that we restructure the media system so that it reconnects with the mass of citizens who in fact comprise 'democracy.' The media reform I envision…can take place only if it is part of a broader political movement to shift power from the few to the many."[25]

To appropriately study the role of communicative media and to evaluate their functioning in a democratic society, a more realistic definition of democracy and a more realistic perspective regarding the role and responsibility of "the press" in this society is needed. This analysis and research generates a new question (that is beyond the scope of this analysis to appropriately answer). It challenges journalism and media scholars, rhetoricians, social and political scientists, and indeed, all of us, to ask and seriously consider: How can we expect our media system to serve the democratic ideals set forth in the American Constitution when those ideals are not evidenced in society?

Writing after the September 11, 2001, attacks on the World Trade Center, journalism scholar and media ecologist James W. Carey asked this question in a different way: "Can we get globalization, democracy, and open communication at the same time or does one of the triad have to be sacrificed to the other—for example, globalization but with a sharp democratic deficit, or enhanced democracy but with necessary restrictions on open travel and communication?"[26] Perhaps an ecological democracy, with citizens responsible for news creation and consumption, and fully conscious of the consequences of their irresponsibility, begins to address Carey's question.

Answers will come collectively from individuals acting in enlightened self-interest to enact critical public policy reform; people using their political power and responsibilities to influence and inform government action. This shift in focus from the problems media inflict on democracy to the problems that a dysfunctional democracy inflicts on the operation of the media brings to light several critical concerns of journalists. Jeff Cohen, executive vice president and editor of the *Houston Chronicle* and former editor of the *Albany Times Union* of New York, believes "newspapers can't be expected to civically engage people. If they're not civically engaged, the delivery vehicle doesn't matter."[27] Well, it does matter, quite simply because everything we do matters! Many news organizations have lost their connection to readers, and readers know it.

More Technology May Mean More but Not Necessarily Better Journalism

Internet and citizen journalism and the new practices, tools, and forms that are emerging are exciting harbingers of a kind of journalism. However, it is naïve to believe that relying on networked and digital news sources alone is sufficient to be informed, and perhaps more importantly, that we will easily transition to a more evolved form of journalism. Using these new digital tools promotes the engagement of readers and citizens, which is a critical component of practicing quality journalism.

One of the most intelligent and articulate voices grieving the demise of the legacy form of news is Alex Jones. Jones made a big media splash with *Losing the News*, released in early 2009 when the next wave of lamenting about the state of the news industry hit. At that time, several big metropolitan newspapers announced the closing or near extinction of their papers. *The New York Times* threatened to close *The Boston Globe*, San Francisco lost its beloved *Chronicle* and in the Midwest, the liberal and well-educated community of Ann Arbor, host to the University of Michigan, made national and international headlines when its 174-year-old *Ann Arbor News* closed and it became the first American city to lose its only daily newspaper.

"The kind of news that I believe in is in trouble," claims Jones, "and this is a problem that must engage us all, as it affects us all."[28] It is truly hard not to believe and join in his nostalgic mourning. Jones is a highly credible, articulate, and well-respected source: His work on *Losing the News* is part of the Institutions of American Democracy series supported by the Annenberg Foundation Trust. Jones worked for *The New York Times* for almost 10 years and won a Pulitzer Prize in 1987 for his story about the collapse of the *Courier-Journal* in Louisville, Kentucky. Jones is also director of Harvard University's Joan Shorenstein Center on the Press, Politics and Public Policy. Jones's lament goes something like this: "What, indeed, is happening to the news at this time of tumultuous technological change? Does it matter that newspapers seem to be in free fall? Is objectivity the best model for American journalism in a new era that prizes the individual voice? Is media concentration a menace or a red herring?"[29]

And while Jones is perhaps nostalgic for "the news he knew," he does quite accurately identify the crisis and articulate it well: "It is a crisis of diminishing quantity and quality, of morale and sense of mission, of value and leadership. And it is taking place in a maelstrom of technological and economic change."[30] It is clear, and has been for a long time, that the newspaper industry is in crisis. In much the same way that the financial, real estate, health care, and automotive industries are experiencing a political, economic, and cultural tsunami of change, the news industry, which has historically adapted well to technological innovation, is again being asked to innovate. The recent shifts and responses—such as the launch of AnnArbor.com by Booth Newspapers the day after the *Ann Arbor News* closed in July 2009 and the 2007 launch of the successful news outlet Politico—are just the latest in the journalism industry's long history of innovation.[31]

The journalism industry's myriad problems have been well articulated by the Nieman Foundation's Reports; the work of Robert McChesney and John Nichols (2010); regularly covered in the blogs of Howard Rheingold, a journalism instructor at Stanford and a longtime Internet expert and community-builder; and NYU professor Jay Rosen. Journalism is truly poised to not only survive but thrive in the age of the newsphere. However, before moving to a close look at the new architecture of news and the potential digital communication technologies provide, it is important to recognize the very real limitations on news and the products of journalism distributed by the Internet.

An industry pioneer and innovator is Arianna Huffington, whose *Huffington Post* was sold to AOL for $315 million in the February 2011.[32] In their comprehensive and timely *Death and Life of American Journalism*, McChesney and Nichols compliment Huffington for the original and breaking news on the site. However, while its commentary is robust (particularly since it gained significant resources from AOL), most of the news that appears on "HuffPost" is still aggregate content from old media. Its small cadre of reporters works hard and get scoops, but they still tend to dig in predictable fields—the White House briefing room, for instance—that downsized print operations are prone to work, according to McChesney and Nichols."[33]

These forward-thinking and tireless advocates for media reform in the United States identify one of the real deficits in the practice of journalism today: the dearth of original, hard-hitting local and regional reporting. "There are not enough HuffPost reporters and they are hard to find beyond the elite warrens of Washington, New York, and Hollywood—although Huffington is now making a sincere and potentially significant effort to highlight blogging from around the country and to get reporters on the ground in such cities as Chicago and New York."[34] Another AOL venture, Patch.com, a network of about 800 hyperlocal news sites across the United States, is building an infrastructure to support local news and community coverage of the 2012 elections.[35]

It is a paradoxical dilemma and real challenge to the industry and its journalists that despite a plentitude of political reporting many citizens do not have regular and accurate access to important information about regional and local elections.

This was powerfully illustrated when the Michigan Truth Squad, a project of the nonpartisan group Truth for Michigan, investigated and reported on the political ads of the Seventh District congressional race in the fall of 2010 (U.S. Rep. Mark Schauer, D-Battle Creek, and former Republican U.S. Rep. Tim Walberg). When the Truth Squad looked at the political ads of these candidates they discovered that each side distorted factual information; the Democrats claimed Walberg missed a decisive vote, which is true, but what the ads didn't say was that he was in the hospital having surgery at the time of the vote. However, the real tragedy goes beyond politicians attacking each other and points to a real void in meaningful coverage on important issues. As John Bebow from Michigan's Truth Squad sadly explained during an NPR interview,[36] there is nowhere voters can go to find out where Walberg and Schauer stand on important national issues, such as the deficit or the war in Afghanistan. Instead it's pure "character assassination."

Something Essential Is Missing in the Newsphere

As the collapse of the Northeast power grid in the summer of 2003 poignantly illustrated, the existence and prevalence of sophisticated technology does not guarantee its proper and reliable function. This is true for the dissemination of electrical energy and for the distribution of news as well. It was especially poignant when the forty-fourth President of the United States, Barack Obama, commented on the crisis in American journalism, weighing in with the ominous prediction: "I am concerned that if the direction of news is all blogosphere, all opinions, with no serious fact-checking, no serious attempts to put stories in context, that what you will end up getting is people shouting at each other across the void but not a lot of mutual understanding."[37] In this respect, Jones is accurate when he laments the loss of meaningful and important news coverage. In 2006, the Project for Excellence in Journalism reported findings from a three-decade study of news coverage in Philadelphia. PEJ concluded that: "There are substantial portions of any given week when fewer than five journalists provide the primary coverage for a city of 1.4 million people."[38]

However, it is both simplistic and erroneous to blame the problems the newspaper and legacy news industries face on the proliferation and growth of the Internet and digital technologies. As we will see next when we look closely at its history of innovation, the journalism industry has consistently (although not easily) adapted well to technological change. And just as the power plant operator ignored the warning signs that shut down the entire network, so too the shortsightedness and greed of many news organizations is largely responsible for their current economic crises. Former *Baltimore Sun* reporter, 2010 MacArthur Fellow, and creator of the HBO series *The Wire*, David Simon explained during Congressional hearings on the state of the industry:

> We know now—because bankruptcy has opened the books—that the *Baltimore Sun* was eliminating its afternoon edition and trimming nearly a hundred reporters and editors in an era when the paper was achieving 37 percent profits.
>
> In short, my industry butchered itself, and we did so at the behest of Wall Street and the same unfettered, free market logic that has proven so disastrous for so many American industries. And the original sin of American newspapering lies, indeed, in going to Wall Street in the first place.[39]

It is important to note here that not all newspapers ravaged their news operations and in keeping with its rich and innovative history, many news organizations such as the *Albany Times Union* carefully and conscientiously adjusted to the digital environment as it built its award-winning TimesUnion.com. Indeed, many realists are suggesting workable solutions and the need for cooperation, experimentation, patience, and flexibility. For example, in their *Saving the News: Toward a National Journalism Strategy*, Victor Pickard, Josh Stearns, and Craig Aaron call for serious policy reform regarding the structural foundations of the American media and news and information delivery structures:

> This is a surmountable crisis, but saving journalism and shoring up democracy's very foundations will require the right application of innovative technology, policy reform, and public resources. There is not a perfect policy solution to solve this crisis. Rather, it will likely be a menu of policy options that together will help fill the vacuum left by the decline of commercial news.[40]

Among the many, many reports, hearings, books, articles, and conferences attempting to shed light on the serious challenges facing the journalism industry and profession was a November 2009 conference sponsored by Yale's Knight Law and Media Program and their Information Society Project. Titled *Journalism and the New Media Ecology: Who Will Pay the Messenger?* the conference explored four key questions about the future of journalism:

1. How will citizens get the information they need to make the informed decisions on which democracy depends?
2. What special values do national and local "legacy" media provide? Should they be preserved, supplemented, or replaced, and if so, how?
3. What role will professionally trained journalists continue to play in gathering and editing news and information?
4. How will those who gather and deliver valuable information necessary for democracy be compensated for their work?[41]

The use of the term "ecology" by conference participants and planners is incomplete and perhaps indicative of the missing link in any real attempts to solve the challenges journalism faces today. The questions the Yale conference posed above (in an attempt to offer solutions to the current crisis) do not account for the reception of news in the environment and the power each individual brings to that process—especially if that action is an evolved and conscious one.

This text is a groundbreaking and progressive application of media ecology theory and its fundamental understanding of communicative media as ecosystems to the practice of journalism and its resulting product—news. Media ecologists are first and foremost media historians who are able to provide the context necessary to understand emergent media and their resulting practices and products. That is why we will next turn to a short explanation of the journalism industry's use of technology.

Journalism's Long History of Innovation

For more than half of its 300-year history, the newspaper industry readily embraced and capitalized on emergent technologies. This was true because publishers and owners found new technology to be a cost-effective means to more efficiently and effectively improve its product. However, in the early 1930s, the first of several new communicative media, radio, created new competition and seriously challenged newspapers as the public's dominant source of news and information. At that point, the newspaper industry's relationship with technology changed. New technologies, especially those that were capable of delivering news and information to the public, challenged the dominance of newspapers as America's first mass medium.

One of the first challenges came from broadcast radio, which was introduced in the 1920s. Much like newspapers, radio stations brought news into people's homes on a daily basis. The introduction of radio caused a great deal of anxiety as well as excitement, and some newspaper publishers initially sponsored some of the first radio stations despite the apparent conflict. In this and other examples of emergent legacy pairs of media, the relationship between the two is usually ambiguous and paradoxical. When faced with competition from radio, the emergent form, newspapers adjusted their reporting styles and conventions: they expanded their stories to include "why," which resulted in the genre of interpretative reporting.[42] Newspapers also used the medium of film to produce newsreels, which were sometimes designed to build audiences. One such example is the Hearst Metrotone News Collection, which is housed in UCLA's film library. Hearst's tabloid style heavily influenced the style and conventions of the newsreels and later television news forms as well.

Newspapers' position as the dominant mass medium in America was soon threatened again by the advent of television in the 1950s and 1960s. At that time, some experts predicted that newspapers would not survive the strong competition posed by this powerful and persuasive new electronic medium. However, desktop publishing and new printing technologies enabled the industry to reduce costs while improving its product through the use of color and graphics. In addition to manufacturing, newspaper publishers found novel ways to use computers in their production. Improved efficiency reduced production time and costs for processes such as pagination. These technological advancements came at a critical time, and they helped the newspaper industry's economic viability, which was a constant struggle given the widespread consolidation effort.

With the rapid introduction of new media, the pace quickened, and newspaper publishers took a proactive stance and funded a variety of research centers to both study and explore the new options technology offered, as well as to protect their economic interests and continued viability. Some of these centers included the MIT Media Lab in Cambridge, which was funded by the Hearst Corporation, and the Knight-Ridder Lab in Colorado, where Roger Fidler, an astrophysicist turned journalist, invented Fidler's Tablet—the first electronic news reader, an early predecessor of the 2010 introduction of Apple's iPad and the 2007 Amazon Kindle. Using research generated at these centers, newspaper publishers were able to launch a series of experiments in computer-based information delivery systems. Some of these ventures were successful, and others were not. One of the largest projects was videotext, the first interactive, consumer, electronic-information delivery system created by French Telecom in 1980. While highly successful in France, American newspaper companies collectively lost approximately $100 million trying to launch similar ventures.[43]

The newspaper industry's next major experiment in online news delivery was Bulletin Board Systems (BBSs), "software packages that enable a personal computer to house a complete interactive online system. The computer, hooked to one or more modems, answers calls from consumers' computers and offers them access to news, information, e-mail, discussion areas, and even teleconferencing in which a group of people can type messages and interact with one another in real time."[44] BBSs were an important transitional step for the newspaper industry because publishers, editors, and reporters learned how to leverage their markets and package online news and information. They were also an introduction to community building online, a key affordance of the Internet, which was recognized as early as 1994 by certain early adopters, such as the WELL, one of the oldest online communities founded in San Francisco.

Then came Mosaic, and everything quickly changed. In January of 1992, Marc Andreessen, an undergraduate student earning $6.85 per hour at the University of Illinois's National Center for Supercomputing Applications (NCSA), designed Mosaic, the first graphical browser for the Internet's World Wide Web. Andreessen's invention, which he made freely available to Internet users, became the "killer application" of the Internet in late 1993 and forever changed the way the world communicates. After Mosaic, the Internet, which originated as the domain of computer-savvy scientists, scholars, and researchers, as well as the fail-safe communication system of the United States government, was now much more accessible to ordinary users. Internet access offered individuals, corporations, and media outlets—especially newspapers—new ways to work, communicate, store, and send and receive news and information.

Once it moved beyond the hands of researchers, government workers, and computer scientists, the Internet was touted as heralding a dramatic revolution in communication. While it is too soon to accurately assess its real impact, media reform activist and communication professor Robert McChesney describes the global computer network as "qualitatively the most radical and sweeping of these new communication technologies." According to McChesney, what distinguishes the Internet from earlier forms is its "all encompassing nature," and he believes it cannot be understood within the context of earlier forms because "it is two-way mass communication; it uses the soon-to-be-universal digital binary code; it is global; and it is quite unclear how, exactly, it is or can be regulated."[45]

Within a short amount of time, a whole new industry emerged to support and capitalize on all phases of the growth and use of the Internet and especially the World Wide Web, which grew at an exponential rate following the introduction of Mosaic. "In May 2001, AOL claimed to have twenty-nine million subscribers. *The New York Times* reported that month that 57 percent of American households had some type of Internet access, and the *Wall Street Journal* said some 461 million people worldwide were connected to the Internet."[46] Some early adopters, such as Stanford University students Jerry Yang and David Filo, eventually became millionaires as investors recognized the monetary value of popular websites, such as Yahoo!, to advertisers and sponsors.

Newspapers Search for Ways to Adapt to Digital Delivery

Similar to the introduction of radio and television as "new media," the Internet with the free, open, and democratic ethos of the World Wide Web, posed a paradoxical challenge to newspaper publishers, who understandably balked at the idea of providing their primary product without charging consumers. For the most part, online editions were generally viewed by newspaper publishers as research experiments, appendages to their main enterprise, and expensive ventures that they were forced to support. Journalism scholar Pablo Boczkowski labels this the "period of hedging," in which innovation was also tempered by "reactive, defensive, and pragmatic traits."[47]

As the Internet's World Wide Web grew in popularity and accessibility, a few newspaper publishers emerged as early pioneers who recognized that online editions were an important communicative medium and not just fragile offspring of print editions. The first newspaper on the Web was the *Palo Alto Weekly*, which launched its website on January 19, 1994.[48] Other early leaders included *Nando. net* (the *Raleigh News & Observer* website), the *Minneapolis Star Tribune*, and the *San Jose Mercury News*, whose archives were available to early AOL subscribers.

By the end of 1995, 330 newspapers in the U.S. were either online or just about to launch a website. In January of 1996, *The New York Times* was on the Web, followed by the *Chicago Tribune*, the *Wall Street Journal*, and *The Washington Post*.[49]

These were truly new pioneers in an emerging new medium. They faced many challenges and constraints, such as the opposition and resistance of a basically conservative industry, the struggle to learn and adapt to new working practices that constantly changed, and above all, to generate revenues. Mistakes were costly in both time and money: they could not afford to invest too heavily in any one area or venture. "Overall, the newspaper industry's involvement with the Internet has been one where it had a lot to lose, and it's been trying not to lose it, as opposed to starting from scratch and having a lot to win," says Steve Yelvington, a former editor of the *Minneapolis Star Tribune* who became a media consultant. According to Yelvington, the newspapers stopped reacting to the dragons of the new-media world. Instead, they are learning to cope with the central premise of the old-media world: make more money than you spend. Yelvington believes the "real strategic challenge to the newspaper industry is that its contents have become largely irrelevant to broad segments of the market."

One outlet that embraced innovation early was the progressive *Albany Times Union*, which with Patti Hart at the helm, adopted a then-unprecedented "Web first" strategy for the metropolitan daily's coverage of the trial of Amadou Diallo, a high-profile international story. It served the news organization well and it remains progressive and financially viable. What is unprecedented now in the history of journalistic innovation is the exponential rate of change. With social networking tools, iPhone apps, Twitter feeds and decks, a whole new language and grammar are being invented to share news and information. As we will see next, navigating the newsphere and using technology well is much more than an economic challenge: it is a moral imperative.

ENDNOTES

1 In subsequent years, a large amount of evidence has proven that Iraq did not have nuclear weapons and was not building them and the invention of their existence was used, in part, to justify an unnecessary war. Since 2004 numerous documentaries, films, books, and other genres have accurately reported this event.

2 The phrase "weapons of mass destruction" was and is used so frequently that it is now popularly recognized by its acronym WMDs.

3 The transcript of Moyers interviewing Lopez in April 2010 is available at http://www.pbs.org/moyers/journal/04302010/transcript3.html. Retrieved May 12, 2011.

4 You can watch the video of Moyers interviewing Lopez at http://www.pbs.org/moyers/journal/04302010/watch3.html. Retrieved February 8, 2011.

5 Moyers, B. (1989). "Image and Reality in America." *The Public Mind*. Public Affairs Television.

6 Please see the previous note. This statement is from Moyers, B. (1989). "Image and Reality in America." *The Public Mind*. Public Affairs Television.

7 Robert Darnton's "Writing News and Telling Stories," which was published in 1975 in the spring issue of *Dædalus*, the journal of the American Academy of Arts and Sciences.

8 Please see the previous note. This quote can be located on p. 176.

9 Darnton, R. p. 181. Darnton's impressive academic and professional credentials include graduation from Harvard in 1960; attendance at Oxford University on a Rhodes scholarship, where he earned a PhD in history in 1964 studying with Richard Cobb (among others). Darnton worked as a reporter for *The New York Times* from 1964 to 1965—the inspiration for "Writing News and Telling Stories" perhaps. He was awarded a MacArthur Fellowship in 1982 and joined the Princeton University faculty in 1968. He currently has emeritus status at Princeton, and was appointed Carl H. Pforzheimer University Professor and Director of the Harvard University Library.

10 Fiske, J. *Reading the Popular*. pp. 149–150.

11 Fiske, J. p. 150.

12 See Fiske, J. p. 193.

13 See Fiske, J. p. 196.

14 See the previous note for Fiske, J. This citation appears on p. 194.

15 Michael Schudson cited in Mindich, D. p. 5.

16 Mindich, D. (1998) *Just the Facts: How "Objectivity" Came to Define American Journalism.* New York: New York University Press.

17 See Mindich, D. pp. 139–143 for these references.

18 Please see Starr, P. pp. 388–389.

19 Please see Starr, P. p. 395.

20 Please see Sinclair, Upton. *The Brass Check: A Study of American Journalism.* p. 429. Sinclair's text is free and available online. I found a copy at chss.montclair.edu/english/furr/hj/sinclairtbc.pdf. Retrieved May 13, 2011.

21 Chomsky believes there is an inherent conflict of interest between the imperatives of a "private, advertising-supported and competitively driven broadcasting system" and the functioning media, who were "driven to find ways to appeal to audiences." Please see *Manufacturing Consent: The Political Economy of the Mass Media*. (1988). pp. 394–395.

22 Please see Carey, J. p. 170. Carey was on the faculty of the Columbia School of Journalism. His opening day speech at Columbia, "The Struggle Against Forgetting," captures the essence of journalism's purpose.

23 Please see McChensey, R. p. 336. McChesney's credentials are impressive. In addition to the publication of eight books on the American media and its relationship to democracy, including *The Death and Life of American Journalism: The Media Revolution That Will Begin the World Again* with John Nichols, in 2001, *Adbusters* magazine named him one of the "Nine Pioneers of Mental Environmentalism" and *Utne Reader* in 2008 listed him as one of

their "50 Visionaries Who Are Changing Your World." Please see his personal website for more detailed biographical information, available at http://www.robertmcchesney.com/.

24 Please see http://www.freepress.net/.

25 McChensey, R. (1999). p. 3.

26 Carey, J. (2006). p. 105.

27 Interview with Jeff Cohen, July 2003.

28 Jones, A. p. xiii.

29 Jones, A. pp. xvii; 178.

30 Jones, A. p. xviii.

31 Please see the following *AJR* article about the January 2007 launch of Politico: http://www.ajr.org/article.asp?id=4265.

32 For more details about the controversial AOL-Huffington Post merger, please see: http://articles.latimes.com/2011/feb/07/business/la-fi-aol-huffington-post-20110207 and http://www.thedailybeast.com/blogs-and-stories/2011-02-07/aol-buys-huffington-post-arianna-huffington-calls-it-her-last-act/.

33 Please see McChesney, R., & Nichols, J. (2010). p. 95.

34 Please see the previous note, particularly pp. 95–96.

35 Please see http://www.huffingtonpost.com/arianna-huffington/local-voices-hyperlocalb_b_857782.html.

36 October 5, 2010, 8:45 a.m.

37 Please see McChesney, R., & Nichols, J. (2010). p. ix.

38 McChesney, R., & Nichols, J. (2010). p. 36.

39 For this reference, please see McChesney, R., & Nichols, J. (2010). p. 37.

40 *Saving the News: Toward a National Journalism Strategy* by Victor Pickard, Josh Stearns, and Craig Aaron. Please see p. 46.

41 This was retrieved May 25, 2010 and is available at http://www.law.yale.edu/intellectual-life/10123.htm.

42 Please see Roger Fidler's *Mediamorphosis: Understanding New Media*. p. 69.

43 Please see the previous reference to Fidler's text. This reference is on p. 43.

44 Fidler, R. p. 45.

45 McChesney, R. p. 121.

46 McChesney, R. p. 52.

47 Boczkowski, P. p. 49.

48 Boczkowski, P. p. 50.

49 Boczkowski, P. p. 51.

· 4 ·

THE ROLE OF PERCEPTION IN THE NEWSPHERE

"This is a fluid universe where what one is looking for determines what one sees."

—Albert Einstein

Neil Postman coined the word *technopoly* to describe the pervasive and mostly unconscious influence of communicative technologies on the American psyche and culture. He warns: "Something has happened in America that is strange and dangerous, and there is only a dull and even stupid awareness of what it is—in part because it has no name. I call it Technopoly."[1] In his seminal *Technopoly: The Surrender of Culture to Technology,* he explains that "technological change is neither additive nor subtractive. It is ecological. A new technology does not add or subtract something. It changes everything."[2] And in sharp contrast to the dominant American cultural narrative that celebrates the ease, efficiency, and benefits of technological innovation, Postman is highly critical and often cynical about its value. "In the American Technopoly, public opinion is a yes or no answer to an unexamined question. An opinion is not a momentary thing but a process of thinking, shaped by the continuous acquisition of knowledge and the activity of questioning, discussion, and debate," says Postman.[3] This definition has far-reaching consequences for the newsphere, which often finds singular opinions elevated to the status of tangible and verifiable truth.

Postman is at times often harsh and consistently unwavering about the real dangers to both individuals and institutions from American's celebratory, devotional, and unexamined use of communicative technology. Writing in 1992—well before the days of Facebook, Twitter, and the constant spawning of social media—he warned that "technology creates a culture without a moral foundation."[4] Telegraphy created context-free information, and the penny press birthed the "sale ability of irrelevant information," according to Postman. And he worried, among other things, about the lack of access to critical information in a culture where those who have power "form a kind of conspiracy against those who have no access to the specialized knowledge made available by the technology."[5]

Faster, further, exponentially more and more, Americans are literally and figuratively drowning in a sea of irrelevancy and fighting a losing battle against waves of news noise: "As the supply of information is no longer controllable, a general breakdown in psychic tranquility and social purpose occurs. Without defenses, people have no way of finding meaning in their experiences, lose their capacity to remember, and have difficulty imagining reasonable futures."[6] It is indeed challenging to know how to assign value to the constantly accessible news and information enveloping the newsphere. The quality and utility of news has been replaced by how much, how far, and how fast. Paradoxically, however, the process of meaning-making is more important than ever. "Most of the information (we get) has little to do with our lives. And most of the time, we don't know what to do with it," writes Postman. And as the amount of information provided increases, its significance and value decrease. Environmentalist Bill McKibben, author of *The Age of Missing Information*, fears that we now live in "an unenlightenment age."

> We believe that we live in the "age of information," that there has been an information "explosion," an information "revolution." While in a certain narrow sense that is the case, in many important ways just the opposite is true. We also live in a moment of deep ignorance, when vital knowledge that humans have always possessed about who we are and where we live seems beyond our reach.[7]

These are complex challenges for which Pulitzer Prize-winning reporter Alex Jones and other experts in the journalism field have yet to discover workable solutions. Like McKibben, Postman poetically articulates the very real harms created by the American technopoly and the pervasive and dire societal consequences of unexamined media reception, but he does not offer to "fix" the problem. As with most ecological challenges, the solution starts with us.

The Individual Is at the Center of the Creation of Integral News

A Wikipedia search for "genres of journalism" renders a surprisingly long list of descriptors for both the historical forms of journalism that have been practiced and consumed, such as *investigative*, *celebrity*, and *new journalism*, as well as several more critical terms, such as *churnalism*. In his *We're All Journalists Now*, First Amendment legal expert Scott Gant asks what may be the proverbial $64,000 question, "So what should we call this new breed of journalism?"—a new genre of journalism that demands both conscious consumption and creation. According to Gant, the long and constantly growing list of current descriptors includes: "Stand-alone Journalism; Grassroots Journalism; Ad Hoc Journalism; Personal Journalism; Bottom-up Journalism; Participatory Journalism; Networked Journalism; Collaborative Journalism; Open Source Journalism; 'We' Journalism; 'We-dia' (contrasted with Media); and Citizen Journalism [the term Grant prefers]."[8]

While citizen journalism address the exciting new creative responses and responsibilities ordinary individuals now possess, it does not fully embrace the connective demands of the newsphere. The adjective that is most congruent with this Teilhardian analysis and the need for a conscious and growing interconnectedness is *integral journalism*, which is built on the premise that current journalistic practices and products now serve to stimulate debate instead dialogue.[9] On his Echo Chamber site, Kent Bye, a documentary filmmaker and former radar systems engineer, claims that the "He Said / She Said" objective standard of mainstream journalism stimulates the polarizing aspects of debate. Its main flaw is that a debate style of news isolates individuals and prevents them from real and meaningful interaction with and exposure to points of view that are different from their own. In its current state, most journalism is not effective at helping people bridge this polarized gap, according to Bye.

Bye's film, *Echo Chamber*, is an open source, investigative documentary about how the television news media uncritically echoed the executive branch's agenda regarding the United States' involvement in Iraq. Bye's use of integral journalism embodies both rhetorical theorist Wayne Booth's concept of listening rhetoric as well as the concept of the "quantum rhetor" who is capable of seeing both sides of an argument simultaneously, a theory I developed with the University of Toronto's Robert K. Logan. Bye can perhaps best be described as a practical integral theorist, who has worked to implement the ideas of Teilhard de Chardin, Ken Wilber, and other philosophers. Unlike a debate style of news, Bye advocates a dialogue approach to the creation, design, and consumption of news, a form that "encourages and stimulates dialogue between opposing viewpoints so that the meaning can be co-created by the collective."[10] In contrast to the current styles of journalism now being practiced, this integral approach to news—a dialogue style—invites and encourages news consumers, readers, and users to explore and fully understand the value of multiple sides of arguments and to, most important, create their own meaning. The following chart explains the differences between debate and dialogue styles of communication. It was adapted from the work of integral philosopher and political scientist Mark Gerzon and serves to guide and assist all levels of journalists in the conscious consumption and creation of an aware newsphere.

DEBATE	DIALOGUE
You have the right answer.	Participants collaborate on an answer.
You prove the other wrong.	Participants collaborate toward a common understanding.
You win.	Participants explore common ground.
You listen to find flaws and make counterarguments.	Participants listen to understand and find meaning and agreement.
You defend your own assumptions as truth.	We reveal and reevaluate assumptions.
You see only two sides of an issue.	We see all sides of an issue.
You defend your view against the views of others.	We acknowledge that others' thinking can improve our own.
You search for flaws in another's position.	We search for strengths and value in others' positions.
You win or lose and close discussion.	We keep the topic open even after the discussion formally ends.
You seek a conclusion that ratifies your position.	We discover new options. There is no closure.

How Do We Better Understand Readers' Perceptions and Misperceptions?

Integral journalism demands a heightened sense of awareness and perception—a genuine openness to the truth and to the reality of one's environment. This is not always easy. Media ecologist extraordinaire, Marshall McLuhan, claimed that in essence, most of us are numbed. How then do we wake up from a "trance imposed on us by our senses," as McLuhan describes the profound media influence on the newsphere and beyond? We start by paying attention to how we see, which will be defined here as perception. Studying and better understanding perception and how human beings see is of critical importance to journalists, media ecologists, and communication scholars now because the processes of perception routinely alter what humans see.[11]

"Perception is the next frontier," Eric McLuhan, Marshall McLuhan's son, told audiences at the 2007 Convention of the Media Ecology Association in Mexico City. Eric McLuhan accurately predicted the next stage of development in the study of human communication and delivery of media messages. The study and understanding of perception is of particular concern to journalists, journalism students, and media scholars. When people view something with a preconceived concept about it, they tend to take their existing notions and "see" them whether or not they are actually there, according to French phenomenologist Maurice Merleau-Ponty.

This problem stems from the fact that humans are unable to understand new information without the inherent bias of their previous knowledge. A person's knowledge creates his or her reality as much as the truth because the human mind can only contemplate that to which it has been exposed. When objects are viewed without understanding, the mind will try to reach for something that it already recognizes in order to process what it is viewing. That which most closely relates to the unfamiliar from our past experiences makes up what we see when we look at things that we don't comprehend.[12] Thus, as you might imagine, the perceptual bias that favors the confirmation of old knowledge over the reception and accurate processing of new information has far-reaching and profound consequences for both the creation and consumption of news in the newsphere: Even when exposed and confronted with facts that are indeed true, people often select and, even more devastating, distort those facts to fit existing belief systems. This behavioral and cognitive phenomenon has been termed "backfire" by political scientists at the University of Michigan.

Lead researcher on a series of studies in 2005 and 2006, Brendan Nyhan explains that backfire is "a natural defense mechanism to avoid cognitive dissonance.[13] Writing in the journal *Political Behavior*, Nyhan and co-author Jason Reifler observe: "An extensive literature addresses citizen ignorance, but very little research focuses on misperceptions."[14] "The backfire effects that we found seem to provide further support for the growing literature showing that citizens engage in 'motivated reasoning.' While our experiments focused on assessing the effectiveness of corrections, the results show that direct factual contradictions can actually strengthen ideologically grounded factual beliefs—an empirical finding with important theoretical implications. Many citizens seem unwilling to revise their beliefs in the face of corrective information, and attempts to correct those mistaken beliefs may only make matters worse. Determining the best way to provide corrective information will advance understanding of how citizens process information and help to strengthen democratic debate and public understanding of the political process."[15]

This is a significant finding because it directly challenges the journalistic credo of objectivity and the call for "just the facts." As Nyhan and Reifler have discovered, even though news stories are reported truthfully and factually, they can be misunderstood and distorted when audiences do not have a pre-existing frame of reference. Their research demonstrates that audiences can easily distort the new information they receive in news stories to fit their individual histories and knowledge no matter how truthful they are.

It is important to make the distinction here between news stories that are readily and easily communicated but challenge existing belief systems and knowledge, and those that are complex and difficult to understand, such as the United States' 2007 to 2010 subprime mortgage crisis. The financial crisis precipitated by adjustable rate mortgages, among other things, was a widely reported and high-impact news story with relevance to many, many audiences. Truly understanding this story required specialized knowledge to unravel and process industry jargon, such as "the shadow banking system," and "credit default swaps," for example. However, NPR's Ira Glass did a stellar job making the complex and murky workings of the mortgage industry and its victims clear and compelling in his "Giant Pool of Money" story, which received wide acclaim.[16]

Nyhan and Reifler's findings are significant because they break down the process of misperception and show how individuals both consciously and unconsciously disconnect from the truth. When the role of perception is considered in the operation of the newsphere, it supports and highlights a simple but profound insight: despite a huge amount of factual evidence to the contrary, people will often deny the truth. Not only will they deny the truth, they will find a way to rationalize factual evidence that conflicts with their belief system and distort that evidence so it confirms their existing worldview. When this is done unconsciously, as it often is, people literally and figuratively don't know what they don't know!

Understanding the role individual perception plays in the interpretation of news and information in the newsphere is of critical importance now because technology can powerfully manipulate attention. For example, many news outlets purposefully ratchet up the level of importance of their stories to attract attention because they base their advertising revenues on the ability to attract eyeballs, click-throughs, and larger and larger audiences. It may appear that simply choosing not to pay attention to certain news stories, outlets, or trends is the solution, but unfortunately it's not that simple because the relationship between attention and distraction is very complex.

In their insightful *Designing Choreographies for the "New Economy of Attention,"* Eric Gordon and David Bogen highlight the very real intricacies inherent in living in a news-saturated world and the paradoxical relationship of attention and distraction. Their work is also an intelligent response to the laments that the contemplative and reflective spaces now created by linear reading will be eroded as reasoned discourse is replaced by brief Twitter-style messages. "Distraction and attention go hand-in-hand," say Gordon and Bogen. "The very same new technologies and landscapes that cultivate a state of distraction are themselves directed simultaneously toward the cultivation of attention."[17]

They redefine attention as "controlled distraction carefully directed towards a particular goal." As a result, they see distraction as not only as necessary as attention, but also as a part of the cognitive processing commonly associated with the practice of "paying attention." Viewed in this context, distraction is as necessary and important to knowledge creation and meaning-making as is paying attention. The critical skill that helps news consumers, and indeed all of us, refine our perceptive abilities and practice conscious consumption is the sophisticated ability to notice distractions and direct our attention with awareness: This is the valuable, higher-level skill now required of news users. As Gordon and Bogen point out, it is important to understand that the relationship between knowledge creation and attention is dynamic, "nuanced," and "nearly absent" from contemporary discussion about how knowledge is processed and the role of each individual's perspective and abilities in that process.[18]

What the research team find disturbing is not the ability for technological devices and communicative media to capture and sustain attention: Gordon and Bogen believe the real cause of the social problems now associated with inattention come from "intensifying technologies of attention," the ability to capture an individual's attention in ever more dynamic and powerful ways. Their solution to this classic dilemma, which has been intensified with ever more sophisticated devices and motivations, is an "intentional choreography" that helps the mind elegantly "wander from the 'focal event' with purposeful guidance" which is often "a gentle nudging instead of a commandeering."[19]

Distraction Can Be Necessary and Valuable: One Hiker's Story

A *news fast* is one way to better understand the paradoxical relationship of attention and distraction. Hiking the Appalachian Trail is, among other things, the ultimate news fast, and often thru-hikers—those who spend months on the 2,144-mile trail that starts in Georgia, ends in Maine, and meanders through 14 states—are sometimes reluctant to re-enter society. Months away from the sometimes incessant and constant barrage of news and information in the newsphere clarifies one's relationship to the news environment and certainly puts it in relief. "When you get off the trail, the 'news' just isn't as important anymore. Once you've achieved some degree of separation from normal media, your perspective changes and the significance of what is reported is relatively diminished," says Jack Magullian, also known by his trail name, "Archaeopterix."

Magullian, an ex-patriate from the state of New Jersey now living in New Zealand, has returned to the United States each year for more than a decade to hike the Appalachian Trail and soak up its wisdom. "The more often that you spend extended periods isolated from the media, the more you realize how little things change. As a result, you tend to focus less and less attention on the overall state of everything and (tend to) seek out new sources for more specific information," says Magullian. "I think I tend to look at everything with an open mind until something either turns me off or stimulates my interest. There's no Walter Cronkite for me: I need to know the motivations behind the story. New Zealand's news sources generate a fraction of the content created by larger outlets, such as the AP, FOX, Reuters, the *Inquirer* or the *Guardian*. I understand where New Zealand news comes from and the slant that's placed on it just as I would any individual source."

Negotiating the complexities of a world readied with "intensifying technologies of attention" without the aid of hiking respite to recalibrate our hearts and minds appears daunting, according to Howard Gardner, who is best known for his work defining multiple intelligences. In his *Five Minds for the Future*, Gardner, a MacArthur Fellow and Director of Harvard's Project Zero, believes that today's world demands individuals who "possess disciplined, synthesizing, creating, respectful and ethical minds." Knowledge creation, processing, and especially meaning-making require "a whole new kind of mind," says Gardner. Further, the world of the future, with its ubiquitous search engines, robots, and other computational devices, will demand capacities that until now have been mere options. To meet this new world on its own terms, we should begin to cultivate these capacities now, according to Gardner.[20]

And, as we so poignantly saw in the story of the breakdown of the power grid when the choice one individual made cast millions in darkness, the mind of the future is inextricably intertwined—it is one mind: "In the inter-connected world in which the vast majority of human beings now live, it is not enough to state what each individual or group needs to survive on its own turf. In the long run, it is not possible for parts of the world to thrive while others remain desperately poor and deeply frustrated. Recalling the words of Benjamin Franklin, 'We must indeed all hang together, or most assuredly, we shall all hang separately,' advises Gardner.

Intelligence is really awareness—the possession of an open mind and the wisdom to use it, as hiker Jack Magullian articulated. It is also an openness, curiosity, and the desire to obtain and process information. Is evolutionary progress like riding an elevator and witnessing the world from one floor, two, or many floors up? Everything looks different when viewed from above; patterns are readily visible and it's easy to observe things that were previously unrecognizable. If this is so, then where is the "up" button and how is it pressed? Cultural anthropologist Mary Catherine Bateson, daughter of the celebrated anthropologists Gregory Bateson and Margaret Mead, believes that making time for self-reflection is the key to possessing "active wisdom."[21] Mary Catherine Bateson, who has followed in her parents' footsteps and is the author of *Composing a Life*, *Willing to Learn*, and numerous other books, believes "we are on the cusp of a change of consciousness about the shape of lives." The primary way people do this is that they learn to "see themselves differently through a process of conversation," according to Bateson.

This is a gradual and historically significant pattern in America, according to Bateson, and she points out that "consciousness raising" movements were the impetus behind the liberation of women and the proliferation of civil rights for minorities. With age does indeed comes wisdom, according to Bateson, and while earlier generations have the benefits of longevity, society is now learning how to cope with incomplete knowledge: "We are all on stage without a script." The problem with this process, as she points out, is that people are not made conscious of their ways of processing information. This is especially true for students, she believes, who need to reflect on their learning.

We all do. "Learning is our primary form of adoption. People in their 60s are skilled at thinking about change," says Bateson, who cautions that there is a real danger when cultures and society focus increasingly on short-term thinking: "Right versus wrong answers in school (especially those presented out of context) lead to adversarial conversations." She believes aggressive media consumption and use have been institutionalized in our education system and the result is less time and support for personal reflection and a "huge tolerance for inconsistency."

The next chapter looks at ways to process information, introduces the "news as a network" concept, and explains how creating a conscious flow of news and information can prevent the short-term thinking Bateson warns against.

ENDNOTES

1 Postman, N. *Technopoly.* p. 20.

2 Postman, N. *Technopoly.* p. 18.

3 Postman, N. *Technopoly.* pp. 134–135.

4 Postman, N. *Technopoly.* p. xii.

5 Postman, N. *Technopoly.* p. 9.

6 Postman, N. *Technopoly.* p. 72.

7 McKibben, B. (1992). *The Age of Missing Information.* New York: Random House. p. 18. Cited in Dery, M.

8 From Gant, S. *We're All Journalists Now.* This citation is on p. 34.

9 I first discovered this concept on Kent Bye's Echo Chamber Project website, which is located at http://echochamberproject.com.

10 Retrieved November 22, 2010, from http://echochamberproject.com. Rhetorician Wayne Booth and French media scholar Pierre Lévy both also claim that intelligence is created collectively. Please see Booth's *Rhetoric of Rhetoric: The Quest for Effective Communication* and Lévy's *Collective Intelligence: Mankind's Emerging World in Cyberspace.*

11 In the second book of the *Reality Transurfing* series, Russian quantum physicist Vadim Ze-land points out the location of power in determining our reality: "The human mind is try-ing to affect the reflection without any luck, while it is the actual image itself that has to be changed. In the end, you get what you believe in." Zeland, Vadim. (2008). *Reality Transfur-ing: 2. A Rustle of Morning Stars.* Winchester, UK: O Books. pp. x; 99.

12 Merleau-Ponty, M. (2002). *Phenomenology of Perception.* (2nd Ed.) (C. Smith, Trans.) Rout-ledge. pp. 484–486.

13 Keohane, J. (2010, July 11). "How Facts Backfire." *The Boston Globe.* http://www.boston.com/bostonglobe/ideas/articles/2010/07/11/how_facts_backfire/.

14 Nyhan, B., & Reifler, J. *Journal of Political Behavior.* p. 303.

15 Please see the above note and reference for Nyhan, B., & Reifler, J. These citations appear on pp. 329–330.

16 http://www.thisamericanlife.org/radio-archives/episode/355/the-giant-pool-of-money. Re-trieved February. 2, 2011.

17 Gordon, E., & Bogen, D. (Spring 2009). "Designing Choreographies for the 'New Econo-my of Attention.'" *Digital Humanities Quarterly, 3*(2). Retrieved February 3, 2010. Endnote: The authors cite Bauerlein, 2007. Cited in Gordon, E., & Bogen, D. p. 1.

18 Gordon, E., & Bogen, D. p. 2.

19 Gordon and Bogen's work builds on Richard Lanham's 2006 *The Economics of Attention: Style and Substance in the Age of Information.* Chicago: University of Chicago Press. Lan-ham's book argues that the ability to generate attention (eyeballs, hits, algorithmically driv-en websites) is the most valued commodity and asset in the culture.

20 Gardner, H. (2006). *Five Minds for the Future.* Boston: Harvard Business School Press. Please see pp. 2 and 9 for these citations.

21 Bateson spoke about her research and the concept of active wisdom in June 2010 at the an-nual convention of the Media Ecology Association at the University of Maine in Orono.

· 5 ·

AN ECOLOGY OF NEWS

"All around us, tangibly and materially, the thinking envelope of the Earth—the Noosphere—is multiplying its internal fibers and tightening its network; and simultaneously its internal temperature is rising, and with this its psychic potential."
—Teilhard de Chardin, The Planetization of Mankind

"Instant access to information changes how people think and act," says Ross Johnson, a successful thirty-something Web designer in Ann Arbor, Michigan. Johnson, who started building websites as a hobby in high school, misses the lively debates with his small cadre of friends, who boisterously jockeyed to become "top dog" on a topic. Instead of comrades, Ross's clan has now become composed of aggressive fact-checkers who quickly whip out their BlackBerries or iPhones and instantly fact-check each other's claims to validate their accuracy. "Now someone will take a stance and someone will take another stance and everyone pulls out their phones and Googles the information and that debate ends fairly quickly because someone finds out this is or isn't a fact," says Johnson.

Savvy users like Johnson and his friends welcome the egalitarian nature of the flow of news and information in the newsphere where, "information gets passed around naturally rather than someone pushing out media constantly," says Johnson. A whole host of bookmarking sites, such as Deli.icio.us, Digg, and Reddit, have grown up around the concept of user popularity (i.e., if you find a site useful, interesting, and informative, you will bookmark it). These collective bookmarks are then aggregated and ranked, and these popularity rankings are used to determine the quality and credibility of the information they contain. These user-driven bookmarking sites are valued for their authority and impartiality, and as a result, prominence and frequency are becoming measures of quality in the newsphere. "It's much more democratic now," says Johnson, who finds more value in information ranked by the people actually looking for it than by editors deciding which stories to publish.

"The more they like it, the more they pass it on, and the more likely it is to show up in search engines and on sites such as Deli.icio.us, Twitter, or Facebook," says Johnson. Users drive story selection in the newsphere. That selection process previously belonged to editors or editorial teams in news organizations. By tracking popular information on social networks, it's easy to see what information people find most relevant and useful, and smart outlets are using this information to drive their editorial strategies, according to Johnson. "[Using data from social networks,] news producers can really see what sort of article works best for the user and allocate resources there," he says.

This emerging ability to self-select stories is also being applied to the choice of broader topics—a vital part of the proverbial news bucket. "Within news as a broad category, there are custom-tailored niches for people who care deeply about specific topics," says Johnson. This is where RSS feeds (Real Simple Syndication) and other push technologies are useful.[1] This is also a place where deep fissures or misinformation can occur (and where professional reporters and editors are still very much needed). Most users are not savvy enough to filter out information that is not useful to them, and they rely on search engines to meet their needs. Determining the quality, relevance, and importance of news, which is much more sophisticated then a popularity contest, is needed now more than ever. (We will look at specific strategies for consuming news in the next chapter.) News consumers must first recognize that there may be important stories that they are missing (i.e., they may not know what they don't know) and then work to create a broad flow of personally valuable and relevant news.

There are a variety of tools that work to filter news through networks, such as a dashboard. "With the growth of mobile [devices] I think your dashboard will be an application built on your phone that serves the same kind of purpose but it is much more compact. You'll be able to get notifications and updates anytime about stories that are important to you," says Johnson. "I get a lot of news following people and organizations on Twitter, such as AnnArbor.com and *Car and Driver* magazine," he says. For Johnson, frequency is one determinant of value. For example, he pays closer and closer attention to a story when it comes to him from multiple sources: "If there's an important story, chances are I'll see it at some point, certainly quicker than if I had to wait for a newspaper to show up because there's often discussion around it. Several people might publish a link to an article. I might see an article or a link to an article posted on Facebook. It might be sent to me as e-mail, too, which is a strong indicator or whether or not I should pay attention to it. If I see two or three people I trust mentioning the same article, then I'm much more likely to look at it even if its headline is not something that would normally have caught my attention," says Johnson.

This is an important extension of the traditional news value of relevance. In the past, reporters and editors determined a story's relevance based on their knowledge and acquaintance with their readers. Audiences selected news sources and outlets because they proved to be useful over time. Relevance now applies much more broadly and immediately to both the specific reader or user and also the user's entire news network—the people they follow on social media and the people who follow them. For Johnson, that number is about 250, which he believes is "not really that many."[2] He follows about 100 other Twitter accounts and uses filters, such as TweetDeck, to segment "tweets" into different groups, which he then ranks in order of importance. "I pay the most attention to the people I know. The important people in my industry are slightly less important and it descends from there," says Johnson, who monitors his Twitter feed throughout the day, much as the previous generation of news users kept a radio on and listened to it while doing other things.

While recognizing the high volume of background noise in his network from Twitter and other sources, Johnson says people who use Twitter a lot are accustomed to doing a lot of filtering of the Twitter stream: They take what they need and leave the rest. Johnson advises making decisions about what kind of news you want to get and identifying credible sources. "If you want to learn about solar energy, for example, find out who the top bloggers are and follow their blog for news because not only are they going to publish their own thoughts they're also going to promote their top sources," says Johnson. He suggests looking for specific Facebook groups and Twitter followers around a topic. "Build a strong network of information, refine it, and keep it coming to you. Find that number one trusted source. Typically, if they are the number one trusted source [on a specific topic] they've gotten there [or at least in the top five] using this democratic process," says Johnson.

The Media Ecology of News

It is exciting to see young news users, such as Ross Johnson, adapt so well to the dramatic changes in the news environment. The next generation of news consumers (and especially those after them), will most likely adapt very well to the evolving newsphere with very little if any recollection of past journalistic practices. However, understanding the fundamental differences between the legacy model of news and today's digital version is critical for both conscious consumption and creation. Before the advent of digital technologies and the many-to-many communicative forms they created, we relied on a hierarchical and industrial model of news. The legacy model was built using analog technology, which until recently, was the dominant means of producing and distributing print and electronic broadcast media. The qualities of that news included centralization, filtering, one-to-many distribution, and profitability. In contrast, an ecological model of news and information delivery uses digital technologies (computer networks supported by the Internet's World Wide Web, primarily) to produce news and information that is decentralized, unfiltered, many-to-many, and egalitarian.

This shift is changing the newsphere and all the news within it quickly and dramatically. These changes altered the media environment as well as in the communicative transactions it supports. These emerging shifts and trends include user-generated content, which in the journalism discipline is often described as the citizen journalism movement.[3] In his progressive and engaging *Watching the Watchdog*, Stephen Cooper celebrates the new communicative forms that grow and expand in the newsphere, such as blogs, which are built around and evolve through the "voluntary interactions and exchanges among people" in contrast to those communication structures that are created "through the deliberate exercise of power, however well-intentioned."[4]

In describing these discourses as "voluntary" and "more beneficial," Cooper has identified a critical component of an ecology of news: the value of individual responsibility in the process. Much like natural selection in the biological world, the internally motivated drive to both search for—consume—as well as create— produce—meaningful and relevant news and information is the crux of an ecological system. Even more to the point, Cooper describes the ideal situation as "Darwinian…the fittest ideas prevail because they are based on the strongest arguments, which are the arguments most persuasive, and hence most acceptable to the participants."[5] To date, most analysis, discussion, and scholarship has focused on either the consumption of news and information in detailed audience and content analysis studies or on the emerging ability of ordinary citizens to create, publish, and distribute content; but these two occurrences have not been viewed as a holistic system. This analysis intends to dissolve the duality inherent in former approaches.

The list below summarizes the systemic differences between the hierarchical model of news and a collaborative, ecological model that features "dialogue style" news.

- "All the news that's fit to print"—Open and evolving stories
- Amplify—Diversify
- Analog—Digital
- Audiences—"Prosumers"
- Broadcast—Narrow-cast
- Centralized—Decentralized
- Commercial—Grassroots
- Common culture—New ideas and alternative perspectives
- Editor-driven environment—User-led environment
- Filtered—Unfiltered
- Fixed—Flexible
- Hierarchical—Heterarchical
- Industrial—Ad hoc
- Inverted pyramid—Spawn discussion
- Market-based—Consensus-based
- One-to-any—Many-to-many
- Paternalistic—Pragmatic
- Product—Process
- Producer-driven—User-driven
- Profitable—Equalitarian
- Proprietary—Community-driven and determined
- Revenue generating—Collaborative
- Self-contained unit—Interconnected
- Snapshot—Comprehensive view
- Suitable for transmission—Context-rich
- Top-down—Participatory[6]

When viewed as an ecology, news is not a product to be consumed but a conscious act to engage and produce shared information that has value in a community: this is how cultures and societies create their histories. Thus news is not an economic transaction but a social and cultural practice involving knowledge generation, information creation, and public distribution. Seeing news as part of the newsphere—a larger ecosystem that incorporates both production and consumption—prioritizes quality over quantity. Consumption of news, as defined here, is a conscious choice necessitating informed thought. It requires the audience to question, and it sheds a different light on the traditional concept of news judgment: It encourages inquiry and it requires participation. By questioning traditional news judgment, audiences can set an alternative agenda and close the loop, if you will, in the consumption-production components of this ecological approach.

When viewed ecologically, the role of filtering and gate-keeping, which had previously belonged only to publishers, editors, and reporters, is now also the responsibility of ordinary citizens. As we will see in the next chapter, news consumers have many tools to use now, such as NewsTrust.net, a social networking–news judgment site that invites members to "think like a journalist," to help them determine the value, credibility, and ultimately the quality of a news story. There are inherent risks in surrendering a practice previously only undertaken by trained professionals, and the historical implications and long-term consequences are yet to be seen. However, in many ways, as Cooper identified earlier (2006), this information is "truer" or more authentic to the audience and would "seem to be…the ideal discourse."[7] As individuals consciously and carefully consider qualities of news and their judgments are duly recorded and quantified, both what constitutes news and the location of the responsibility for the exercise of news judgment will change. As more enlightened news audiences—now fluent in the language of the two-way digital environment—exercise their judgment, we will see shifts in agenda setting, framing, and story selection. Quite simply, who decides what is news will change, and in fact, is changing right now.

How will this change be evidenced? How will individuals affect existing and emerging news outlets and production centers? What new consumption patterns will be seen? These are appropriate and progressive questions, and it is evident that a simultaneous focus on both consumption and production—an ecological approach to news—allows us to open up public discourse to the collective level in new ways. Innovative social networking news sites, such as NewsTrust.net, Digg, Reddit, Deli.icio.us, and others, are shifting the conventional consumption of news. And the production side evolves daily: as I write this, Fox News just launched its MyNews, the 2011 incarnation of the tablet newspaper Roger Fidler envisioned in the 1990s.[8]

Imagine: We're All Journalists Now

"Imagine a world, one easily conceivable today," Michael Schudson, one of the nation's premier journalism scholars wrote in 1995, "[where] each of us [is] our own journalist."[9] Well, that day has come—it is here now—and as Scott Gant so eloquently titled in his recent treatise, *We're All Journalists Now: The Transformation of the Press and Reshaping of the Law in the Internet Age*, citizen journalists are here to stay and ready to assume their new responsibilities. Gant is much more than a cheerleader for the blogging and the fact-checking that helps to keep old media in check. He believes that "the future of journalism depends not only on how suppliers of news organize themselves, but also on consumers. If people do not demand (and pay for) high-quality news and analysis, we are unlikely to get much of it."[10]

"The best journalism is as good as ever," writes Gant. "But we need more of it if the new journalism landscape is going to tell us more of what we should know."[11] So while we are all journalists now to some degree of responsibility and culpability, precisely who writes the stories we read and determine to be news truly matters more than ever.[12] Schudson built his stature in the field with his 1978 *Discovering the News: A Social History of American Newspapers*, which is based on his doctoral dissertation in the Department of Sociology at Harvard. In his book, Schudson develops the history of the ideal of objectivity in American journalism and notes that "the 1830s marked a revolution in American journalism. That revolution led the triumph of 'news' over the editorial and 'facts' over opinion, a change which was shaped by the expansion of democracy and the market, and which would lead, in time, to the journalist's uneasy allegiance to objectivity."[13]

Later, in his 1995 *Power of News*, a collection of essays, Schudson attempts to clarify and elucidate the constant tension created by the conflicting goals of journalism's higher calling as the nation's fourth estate, the highly imperfect translation of that mission by an often greedy media industry, and the reception and consumption of news, which he reminds us "becomes the property of all of us."[14] He argues here that news is, indeed, public knowledge and perhaps that is its organic claim to preserving democracy:

> The news serves a vital democratic function whether in a given instance anyone out there is listening or not. The news constructs a symbolic world that has a kind of priority, a certification of legitimate importance. And that symbolic world, putatively and practically, in its easy availability, in its cheap, quotidian, throw-away material form, becomes the property of all of us. That is a lesson in democracy in itself. It makes the news a resource when people are ready to take political action, whether those people are ordinary citizens or lobbyists, leaders of social movements or federal judges. This is the necessity and the promise of the public knowledge we call news and the political culture of which it is an essential part. (p. 33)

In addition to a reminder about what news does, Schudson points out the news-sphere is part of our collective consciousness: "news is the background through which and with which people think."[15] That is why it matters so much who decides what is news. As journalism historians and their students know, the American press was from its infancy highly partisan and as Schudson points out, "the quest for objectivity itself…is a source of bias."[16] He identifies and explains four kinds of bias inherently imposed by the constraints of the professional practice of journalism in America that result in news that is typically "negative, detached, technical, and official." These trends include:

1. News tends to be bad news. Professional news tends to emphasize conflict, dissension, and battle: out of a journalistic convention that there

are two sides to a story, news heightens the appearance of conflict even in instances of relative consensus.

2. News tends to be detached. Journalists take a distanced, even ironic stance toward political life…that tutors readers in the cool, professional gaze… moving them to stay at home (instead of voting).

3. News emphasizes strategy and tactics, political technique rather than policy outcome, the mechanical rather than the ideological. Focusing on the technical enables the journalist to be professional because then he or she can remain apart from "the conflicts of interest, perspective, and value that are dangerous stuff of political life."

4. News is official, dependent on legitimate public sources, usually high-placed government officials and a relatively small number of reliable sources. News is as much a product of sources as of journalists; indeed, most analysts agree that sources have the upper hand.[17]

Can and should the news that is consumed, produced, and circulated by citizen journalists be different from the professional kind that has been practiced for centuries? Yes, but precisely how that form of news will evolve is impossible to predict. Indeed, it is a transitional time in the history of American journalism: Many things are already dramatically different, including the information structures that support the news and information we receive 24/7. We have many reasons for optimism and an improved news product. Scott Gant explains: "I am hopeful, even though it is hard to imagine exactly how the transformation of journalism will unfold. Will we look back in ten years (or longer!!!) and view as quaint the way news is generated and consumed today? Remember life before e-mail—less than a decade for most of us?"[18]

Theories Behind the Citizen Journalism Movement

Just as digital technologies are creating revolutionary effects, the penny press of the 1830s similarly signaled the inauguration of a commercial revolution in the practice of journalism as well as the rise of news-making as an industry and a business enterprise. This understanding is critical because it helps address some of the structural problems Robert McChesney (1999) identifies in the operation of the American media: "If we value democracy, it is imperative that we restructure the media system so that it reconnects with the mass of citizens who in fact comprise 'democracy.' The media reform I envision…can take place only if it is part of a broader political movement to shift power from the few to the many." This shift McChesney advocates is indeed happening now, and it is often described as the citizen journalism movement.[19]

In an ironic and paradoxical twist, technological affordances, which previously allowed the mass distribution of news and information to large, increasingly homogeneous audiences, are now giving that same power to individuals. Low barriers to entry provided by the Internet and computer networking technologies, as well as new genres, such as blogs, offer a new media landscape for twenty-first-century journalists. Freed from large investments in distribution and production equipment (known as the long tail in marketing terms), individuals and grassroots organizations are pioneering a host of new journalistic styles and practices and generating new communicative media forms, such as YouTube and hyper-local geographically based websites, as well as refreshing older forms, such as obituaries. It is readily apparent that with these new two-way digital tools, the public can easily and cost-effectively produce news. However, an ecology of news demands more: it challenges the public to view themselves as both producers and consumers of news—to view the newsphere as an ecosystem. Production and consumption happen systematically. In addition to consciously consuming information, citizen journalists are not only informed but also engaged and energized to act on their newly acquired knowledge.

One of the most clearly and fully articulated discussions and explanations of the citizen journalism movement is found in the Winter 2005 issue of Harvard's *Nieman Reports*. An excellent summary of the current status of the movement is found in Shayne Bowman and Chris Willis's essay, *"The Future Is Here, but Do News Media Companies See It?"* In addition to lessons learned from successful citizen media efforts and a very important graphic, "The Emerging Media Ecosystem," Bowman and Willis explain what citizen journalists actually do: "Citizens everywhere are getting together via the Internet in unprecedented ways to set the agenda for news, to inform each other about hyper-local and global issues, and to create new services in a connected, always-on society."[20]

The news that citizen journalists choose to share is intrinsically different from the news professional journalists have been trained to report. Thus, it is more important than ever to ask the following questions when studying emergent journalism practices, such as the citizen journalism trend:

- Which conventions from legacy media are being adopted?
- What journalistic practices are being used?
- What new communicative forms are emerging?
- Which of these new forms are medium specific?
- How do these forms work rhetorically? Are they effective?

As an apparently never-ending succession of innovative websites and news delivery systems, such as Wikinews, Google News, and Indymedia, demonstrate, the ability for citizen journalists to seize the potential of digital technology and create novel and effective ways to deliver news and information is unprecedented. In addition to embracing the power and potential of producing news, citizen journalists must also embrace their responsibilities as consumers of news: Rarely are the two seen as an organic and mutually constitutive process. This can be traced, in part, to the fact that the citizen journalism trend and its resulting "products" are still often viewed through the framework of legacy and mainstream media (MSM). For example, in "*Journalism as a Conversation*" (2005), Jean K. Min, director of OhmyNews International, says, "We believe bloggers can work better with professional assistance from trained journalists. On the other hand, we also believe professional journalists can expand their view and scope greatly with fresh input from citizen reports."[21]

Bowman and Willis (2005) also use legacy media to frame their discussion of emergent media in the accompanying text to their graphic, "The Emerging Media Ecosystem." They state:

> The relationship between citizen media and mainstream media is symbiotic. Information communities and weblogs discuss and extend the stories created by mainstream media. These communities and the blogosphere also produce citizen journalism, grassroots reporting, eyewitness accounts, annotative reporting, commentary analysis, watchdogging and fact-checking, which the mainstream media feed upon, developing them as a pool of tips, sources, and story ideas.[22]

The BBC's Director of World Service and Global News division remarked in the *Nieman Reports* (Winter 2005) that "we don't own the news anymore." These remarks and examples show that even the most robust citizen media formats are still often framed within the MSM and legacy models. What is now needed is an ecological approach, which includes the symbiosis Bowman and Willis identify. Media ecologists, as well as rhetorical theorists and journalism scholars, are beginning to understand and write about the importance of using systemic (ecological) approaches to better understand communicative transactions, such as news.[23]

Open Content Is Self-Vetting; Quality in and as the Network

Journalism is undergoing a renaissance as modern-day Michaelangelos and da Vincis—today's digital artists—shape the future of news and information delivery and the political, social, and cultural institutions that result from these new ways of generating knowledge and chronicling history. Freed from the constraints and limitations of print, newspapers, and a one-to-many communication model, news stories are no longer fixed. The unfolding of a news story is a process now, a cultural meme that, like DNA, evolves in the 24/7 newsphere where it is subject to "survival of the fittest" rules. These inevitable evolutionary forces of change keep the news environment open and fundamentally self-vetting. An open environment is structured to produce the truth, something we can all claim as our collective intelligence.

One of the world's leading experts on the applicability and practicality of smart networks is Judy Breck, an educational consultant and blogger on emerging literacies. In her work in the educational environment, Breck has seen firsthand how knowledge works in the new world of the Internet, which she describes as Tim Berners-Lee's "One Web." Knowledge does very different things than it did before: It is connective and its movement creates new patterns. "Open content vets itself. The source is irrelevant," says Breck. "When you take a piece of knowledge and place it in a node in the network, it does different things," she explains. "Findability" is the key to knowledge generation, wisdom, and learning because information is now unbundled and miscellaneous. "When it's open, it can connect. Patterns can arise, meaning is acquired, and that is going to be the new value, that is going to be the new sustainability," says Breck.[24]

Qualities previously attributed specifically to news stories, such as accuracy, balance, and transparency, for example, must now be applied to the networks that distribute and circulate news. News stories now act as cultural memes: Their value evolves as they move through networks and create dynamic connections. Thus news stories are no longer fixed—permanently housed on the pages of a newspaper, magazine, or in a film clip. News may still start as a story, but it now changes minute by minute as it moves throughout the newsphere.

News stories have always evolved and changed, e.g., they are updated and followed. This is not a new process. However, what is groundbreaking now is the way news stories change in the digital newsphere; news stories evolve in a dynamic, real-time way in response to audience reception and use. This changes how quality is determined. Audiences imbue news stories with value based on their relevance and truth, and that value is evidenced in wide (or diminished) circulation and the real dialogue they engender.

This is dramatically different in structure and function from the existence and operation of a news story in the legacy MSM environment. Previously, news stories were packaged and sold as static, fixed entities and their value was determined in large part by their source. Readers read and believed these stories because they trusted and valued the source of the information—that was the primary location of value. Now that news consumers create and manage their own networks, the value of news stories resides in the channels that distribute and proliferate them. Ideally, irrelevant and erroneous stories do not survive the strength of the system. But just as viruses invaded our bodies and our computer networks, we must learn how to vaccinate ourselves against falsehood and manipulation, which means a heightened awareness of source credibility and motivation as well as relevance and impact.

Given the character, structure, and operation of the newsphere, it is critical now to consciously monitor the flow of news and especially its flow toward and away from us. So if news stories were fair, balanced, accurate, well-sourced, enterprising, responsible, original, transparent, insightful, and informative, then so too must be the networks that deliver these stories. It is time to now think of news stories as nodes on the network (or cultural memes, as Richard Dawkins and Tracy-Logan specify) and the networks as the helix or dynamic, ever-evolving structure that supports their birth, life, and death. However, unlike the more fixed and permanent networks with which we are familiar—computer networks, organizational flowcharts, interstate highways, and other transportation systems, such as the London and New York City subway systems—digital news networks are dynamic structures. An example of this dynamic impermanence is the flash mob, which Clay Shirky insightfully describes in his *Here Comes Everybody: The Power of Organizing Without Organizations.*

It is important to be aware of the impermanence of these networks and the advantages and disadvantages of this transience. Unlike print newspapers (and their supporting hierarchical organizations and editorial staffs), the news traveling through these dynamic networks is powered by both human and artificial intelligence—humans that formally rank stories on sites such as Digg or NewsTrust.net, which offers the promise of quality journalism, as well as bots, spiders, and other inanimate creatures that roam the Internet and build a fluid and dynamic ranking—what you see on Yahoo! or Google news, for example. It is also important to be aware of the "filter bubble," an algorithmic editing of information based on historic preferences and to maintain a balanced information diet.[25]

Because news generation, verification, and distribution is a dynamic process that involves exploration, participation, interpretation, discussion, evaluation, and deliberation, a news network is very similar to a newspaper—the place to go to find out what is new, what people in the community—all communities far and wide—are talking and thinking about. Just like a newspaper, news networks update unfolding stories, monitor special interests, and offer occasional pleasant surprises and yes, entertainment, too.

News has always been a choice for those who create and consume it, "an extraction process, saying that one event is more meaningful than the other," according to the *Atlantic's* media analyst James Fallows. In the past, reporters and editors made the choice as they almost exclusively exercised their news judgment and served as the "eyes and ears" around the globe and across the streets. Now, we have the opportunity and consequently, new responsibilities for the exercise of that judgment, and will look next at some specific strategies to do that well.

ENDNOTES

1 RSS is a formatting style on the Web used to publish information that is frequently updated, such as blogs entries, news headlines, and video clips. It is sometimes referred to as a channel and can include summaries, publication dates, and bylines.

2 I interviewed Ross Johnson using Skype on May 2, 2010.

3 Many insightful scholars are breaking new ground in the area of user- and citizen-generated content, including Bowman, S., & Willis, C. (2005); Bye, K. (2006); Cooper, S. (2006); and Hiler, J. (2002).

4 Cooper, S. (2006). p. 302.

5 Cooper, S. p. 279.

6 Additional qualities of the newsphere include: distributed, innovative, adaptive, low-cost, non-proprietary, non-market, hybrid, pro-am, networked, and emerging.

7 Cooper, S. (2006). p. 32.

8 Retrieved February 9, 2011. http://www.innovationsinnewspapers.com/index.php/tag/information-design-lab/.

9 Schudson, M. (1995). *The Power of News*. Cambridge, MA: Harvard University Press. p. 1.

10 Gant, S. *We're All Journalists Now: The Transformation of the Press and Reshaping of the Law in the Internet Age*. p. 199.

11 Gant, S. p. 198.

12 Schudson, M. *The Power of News*. p. 8.

13 Schudson, M. *Discovering the News: A Social History of American Newspapers*. p. 14.

14 Schudson, M. *The Power of News*. p. 33.

15 Schudson, M. *The Power of News*. p. 16.

16 From Schudson, M. *The Power of News*. p. 9.

17 Schudson, M. *The Power of News*. p. 9.

18 Gant, S. p. 198.

19 McChesney, R. (1999). p. 3.

20 Bowman, S., & Willis, C. *Nieman Reports*. p. 6.

21 Min, J. K. "Journalism as a Conversation." p. 18.

22 Bowman, S., & Willis, C. (2005). p. 7.

23 In addition to the work of media ecologists such as James Carey, Walter Ong, and Neil Postman, who recognized the power and potential of emergent media forms, their dangers, as well as their capacity to be shaped with humanism, other scholars are helping to define the ecology of news. Sociologist Kathleen Carley and communication theorist David Kaufer eloquently reinforce the need for an ecological approach to studying communicative forms such as news. According to Kaufer, D., & Carley, K. (1993), "without a systematic ecological perspective…the impact of communication technologies are often misunderstood" (p. 88). They then go on to explain how such an ecology works: "Content, context, agents, and the communicative transaction are inextricably bound into a single ecological system such that affecting one ultimately effects all" (p. 88). Finally, they submit a lens in which to frame the present and continuing study of this process:

> Despite a growing acceptance in the literature that individuals, social structure, culture, technology, and language are somehow related as mutually defining elements, the literature has mainly been silent on positing a specific mechanism tying them together. [What is now needed is] an operational model of communication that is sufficiently detailed or precise enough to permit formal analysis. (p. 206)

Kaufer and Carley have closely analyzed the communicative transaction process, and their definition offers a launching point to build an ecological model of news and information delivery. In *Communicating at a Distance*, they illustrate the role of concurrence within the communicative transaction:

> …The communicative transaction takes place within an ecology consisting of not only concurrent transactions, but their content, context, and agents. Individuals adapt during a transaction, and because of the reciprocity between interaction and cognition, such adaptations lead not only to new mental models but to new socio-cultural positions (and hence roles). Through concurrent and recurrent transactions, changes at the level of the individual collectively construct social and cultural changes. In response to interactive-cognitive reciprocity at the individual level, social structure and culture co-evolve. (Kaufer, D., & Carley, K., 1993, p. 160)

This definition clearly locates the communicative transaction as the interface between the individual and the larger environment. The key components of this interface are content, context, and agents: these components are useful to construct an individual's news transaction.

24 Taken from an interview on her Golden Swamp website. Other relevant work includes Stephen Gray's publications on open-computing and John Seely Brown's "Re-Imagining Dewey for the 21st Century: Learning in/for the Digital Age." Breck's video is available at http://www.goldenswamp.com/about-goldenswamp-2/.

25 See Pariser, E. (2011). *The Filter Bubble: What the Internet Is Hiding from You*. New York: The Penguin Press.

· 6 ·

HOW TO CONSUME NEWS

"One could say that the whole of life lies in seeing....To try to see more and to see better is not, therefore, just a fantasy, curiosity, or a luxury. See or perish. This is the situation imposed on every element of the universe by the mysterious gift of existence. And thus, to a higher degree, this is the human condition."
Teilhard de Chardin, The Human Phenomenon

Sarah Merion, a twenty-something living and working in Boston, has a dynamic relationship with the newsphere. "It's one thing to go seek it out and be proactive, but it's another thing to place yourself in a position to be able to receive news, too. If you're open to different outlets and most of these outlets are online, then you are able to receive information in different manners. How I get news doesn't matter—it just matters that I get it," says Sarah, who combines Twitter and RSS feeds with e-mail, websites, blogs, and e-newsletters (and occasionally word-of-mouth and a little television) to round out her personal news network.[1]

"If it's 'big' news, I go straight to the *NY Times* because I know they have the latest story and it's quick and easy for me to find," says Sarah, who doesn't own a radio and who rarely watches American television. Sarah's news network combines traditional and mainstream outlets, such as *The New York Times* and occasionally CNN with a healthy supplement of links and sources supplied by her numerous Twitter followers, whom she describes as "interesting, insightful, and intelligent."

"My interests are wide, but on Twitter I follow topics related to technology, start-ups, venture capital, and marketing. I also follow a few people who tweet about SEO (Search Engine Optimization) because I really have no idea about it," says Sarah. She also has a category of followers called "people who do incredible things," that she is building relationships with so she can eventually meet them face-to-face.

Getting and staying informed is a constant negotiation, and like most of us, Sarah asks herself, "What should I know about?"; "What is valuable?"; "What and how much should I read?" and similar questions. Being informed means knowing what is happening "in MY world RIGHT now," says Sarah, who fully appreciates the power and immediacy of the newsphere and the ability to follow breaking events, such as the tweets that provided a glimpse into the unfolding of international protest movements. Like most of us, Sarah is also challenged by "information chaos," the term Media Ecology founder Neil Postman used to describe the difficulty in making sense of too much irrelevant information.[2]

It is very easy to access news and information but much more difficult to regularly receive information of value. "We are often shielded in the United States from what really happens in the rest of the world," says Sarah, who has spent several years studying, living, and traveling extensively in South America. Sarah regularly reads the *International Tribune* and consults the BBC to help her achieve a more balanced news diet. "I read a lot of blogs but was feeling irresponsible and almost promiscuous (about) my blog reading! I was at a saturation point where what was supposed to be educational and informational became clutter," says Sarah. She then launched a filter and sorting process to discover what was important and what to forgo. "If you aren't careful and you don't realize it, you'll wake up one day and realize you're reading [garbage] that doesn't mean anything. I'd love it if someone else did the filtering for me, but ultimately I have to filter [the newsphere] myself," she says.

Paradoxically, the only way to regulate the flow of news and information—design a filter—is to determine quality on both a collective and a personal level. It is important to determine what you're missing from regularly accessed news outlets, and the only way to do that is to look critically at multiple sources and outlets. You need to find out what your sources are not telling you, a task that requires a critical approach to the current news and information you are currently consuming. In essence, you need to know what you don't know.

Like fashion and food, news stories also have a distinct style. "I know where I like to get my news and I like the style of certain outlets—the way stories are written. [I believe] knowing what outlets you like and testing those outlets is important because you absorb more," says Sarah. For example, Sarah checks CNN for headline information—"I'm on CNN right now just reading the headlines on the front page to see what's up in the world and if I'm missing anything big," wrote Sarah from Iguazu Falls as she traveled throughout Argentina and Brazil. "The *NY Times* is more my style—more editorial, a bit more creative, and they have a wide array of content that appeals to me." Initially, when setting up a network, it is important to focus on inclusivity—reading and searching widely especially for divergent opinions. Later, after monitoring outlets, it is possible to begin to eliminate sources that are not comfortable, credible, or useful.

Audiences Use Relevance to Create Meaning and Add Value

Sarah Merion is successfully navigating the newsphere because she understands that networks are dynamic entities. To accommodate 24/7 news cycles and almost constant change, she regularly updates and fine-tunes the personal filters that help separate irrelevant stories from those that have real meaning and value. This dynamic quality is also evidenced in the specific stories that make up the news environment. Today's networked news story is an artistic hybrid—a combination of legacy forms and qualities as well as emerging characteristics and styles. Quite simply, some things will stay the same and others will change; it is impossible now to predict. Here is a comprehensive list of the traditional qualities of news as outlined by NewsTrust.net, a website dedicated to discovering quality journalism.

1. Fairness: Is the story based on facts or opinions?
2. Well sourced: Can the story be confirmed by multiple sources?
3. Context: Does the story give the big picture?
4. In-depth: Is it well researched?
5. Enterprise: Does it show initiative? Courage?
6. Relevance: Is the story newsworthy? Is it meaningful?
7. Style: Is it well written? Is it clear? Concise? Compelling?
8. Accuracy: Can you confirm that it is true?
9. Balanced: Does it present divergent viewpoints?
10. Expertise: Are the sources qualified? Knowledgeable?
11. Informative: Did you learn something new?
12. Insightful: Is it well reasoned? Thoughtful?
13. Original: Does it offer a new perspective or fact?
14. Responsible: Are claims valid, ethical, unbiased?
15. Transparent: Are there enough links and references?

These measures of quality originated in the legacy, two-way media environment that saw consumers as primarily receivers of news. And as the fabric of the legacy form of news unravels, a new qualitative tapestry is emerging. Journalistic entrepreneur, blogger, and the brains behind the clever Blue Nile marketing efforts, Ben Elowitz, actively dismantles not only objectivity but the other stalwart standards of legacy newspaper reporting, including credentials (credibility), correctness (accuracy), objectivity (nuetrality), and craftsmanship (inverted pyramid and other legacy forms): "If old-media traditionalists can be relied on for one thing as the world digitizes, it's to bemoan the loss of what they call 'quality.' In fact, the quality of published content has never been better," says Elowitz in a Leading Voice blog post.

The dynamic, two-way structure of the newsphere and the ability for audiences to monitor, distribute, and redistribute news has invalidated traditional qualitative measures, according to Elowitz. "The audience doesn't care where the content comes from as long as it meets their needs," says Elowitz, and he sees referral endorsements of friends and search algorithms replacing authority with relevance. "In the abundant world of content, consumers know how to apply their own sniff tests—and with myriad sources, they develop their own loyalties and reputations. The brand's stamp isn't the point anymore—the consumer's nose is." The ability to constantly monitor the progress and evolution of a story in the newsphere is also altering traditional professional practices, such as fact-checking and error correction. Expert readers and bloggers are now often editing and updating stories to improve their accuracy.

Elowitz's point of view is sometimes extreme and his warnings dire—"publishers that don't move beyond these anachronistic measures of success [i.e., using credentials, correctness, objectivity and craftsmanship to determine quality] will perish."[3] But he accurately describes the dramatic shift in how consumers use news and information in a two-way communication environment: "Looking at these four old criteria for quality [credential, correctness, objectivity, and craftsmanship], they all share the same source: they are based on the belief that a publisher controls the audience's experience; and the audience's access to content is scarce." Now that audiences are judging quality multiple times a day, in multiple ways, and for a multitude of stories, the primary determinant of news quality is relevance, quickly followed by the ability to "make experiences, not content," the reporter or developer's point of view, and finally, the system of distribution of the news (i.e., its network).

Here are some specific qualities of news in the emerging newsphere:

1. Relevance: Do audiences find meaning in the message?
2. Engagement: Is the story memorable? Does it prompt action?
3. Voice: Is the reporter's point of view transparent and credible?
4. Networked: Is there a consistent and open news flow?

A paradoxical reality of the newsphere that presents a clear challenge to its creation and use is the relationship between news about news, which we saw when we looked at *Dateline NBC's* coverage of the Captain Phillips rescue, and the need to locate quality news that has value and relevance to audiences. As a variety of researchers from cognitive scientists to political scientists have demonstrated, readers and users only truly understand and integrate news and information when it makes a genuine connection to their knowledge base and their reality. The high volume of noise in the newsphere is largely irrelevant to a majority of audiences.

However, it's not as simple as going off into the wilderness and hiking the Appalachian Trail to get away from it all, which is something we all often long to do. A powerful wild card in all of this is the unpredictability and, yes, real excitement of a living in a fast-changing world. Who knows what will happen next? We don't know, but we want to know, and so we stay connected. And we stay connected so we can get news and information quickly.

In the newsphere, *fast* now means instantaneous. Here's just one example: Michael Jackson died at 2:26 p.m. PST on June 25, 2009. Time Warner's celebrity gossip site, TMZ, scooped every other news outlet and was the first to report his death at 2:44 p.m. Because of its tabloid style and use of paparazzi, many national and international outlets waited for the *Los Angeles Times's* website report at 2:51 p.m. to then report the story in their own outlets. In an interview with *The New York Times*, TMZ's founder Harry Levin said, "We work as hard at breaking a Britney Spears story as NBC would work on breaking a President Bush piece." Levin, a former investigative reporter for KCBS-TV in Los Angeles, covered the O.J. Simpson trial. He defends his work and the organization he has built: "We've become like The Associated Press in the world we cover," he told *The New York Times*.[4]

More and more breaking news is being reported by a new genre of reporters—the citizen journalists. Janis Krums tweeted one of the first images of the January 2009 landing of a US Airways flight in the Hudson River in New York City: "There's a plane in the Hudson. I'm on the ferry going to pick up the people. Crazy," Krums wrote as she watched Flight 1569's passengers scramble to safety on the plane's wings.[5] Likewise, Stefanie Gordon from Hoboken, New Jersey, captured a breathtaking image of the May 17, 2011, launch of the Space Shuttle Endeavor on her iPhone while on an airplane flight. Her video went viral and quickly brought Stefanie a brief moment of fame.[6]

One of the industry's pioneers and true visionaries is Jeff Jarvis. He writes a popular blog, BuzzMachine, and is a journalism professor and one of the first reporters to blog the 9/11 collapse of the Twin Towers in New York City. Jarvis describes relevance as a "nesting of evidence" that readers build based on a quality flow of information from credible sources. In his *What Would Google Do?*, Jarvis maps relevance against timeliness in a grid that includes brands, search, human links, and predictive algorithmic authority to help explain how individuals select and receive news stories in an environment where the value of brands is diminishing. In a BuzzMachine blog post, Jarvis asks us to:

> ...Imagine a content ecosystem where users—who already do most of the information sharing themselves—decide where and how value can be added, explicitly or through our usage data. Imagine that these creations come to us through recommendations from peers we trust, prioritized by formulae (human-aided algorithms and algorithmically aided humans), or through search. Imagine that the creation isn't a static piece of content but instead that nest of relevance with updates powered by collaboration with links. News and media start to look very different.[7]

These differences in the digitally-driven newsphere are easier to notice in news products, such as TMZ's reporting of Michael Jackson's death, than in actual journalistic practices. As more news producers and consumers grow in confidence and comfort, and expand their community base and networks, the potential for Jarvis's "nests of relevance" will increase.

Strategies for Consciously Consuming News

As you turn on the television, log on to your computer, or connect to the Web with your BlackBerry, iPhone, or iPad and otherwise tap into the newsphere, please first ask yourself as you start to read, watch, and listen to news stories, "Is this really important to me?" Take a moment to decide whether it is in your best interest to pay attention to a particular story. It is the job of the journalist to attract your attention. There are many old and new strategies for doing that. It is your job to figure out precisely what to pay attention to; that is what it means to be aware and act consciously. To fully participate in the growing collective consciousness and create the unity foretold by Teilhard's noosphere, it is critical that each individual wake up and responsibly use his or her energy and power.

Consciously negotiating the newsphere means waking up and recognizing when our attention and energy are being manipulated. "It often happens that what stares us in the face is the most difficult to perceive," Teilhard said.[8] He believed the main problem was "energy leakage: We will struggle to compensate for our imbalance by materialism or an ever-increasing multiplicity of experiments."[9] To help us wake up and stay in balance, four of the savviest authorities on emerging news and digital literacies, Neil Postman, W. James Potter, John McManus, and Howard Rheingold, offer the following strategies.

Neil Postman Explains the Role of Education
in Creating Informed News Consumers

Neil Postman is the founder of the media ecology tradition and the author of *Technopoly*; *Amusing Ourselves to Death*; *The Disappearance of Childhood*; and other important texts. Neil Postman was a humanist, cultural critic, and the founding father of media ecology as an academic discipline. He championed education as the anecdote for the predominantly unconscious societal effects of technology. He wisely remarked: "I don't think any of us can do much about the rapid growth of new technology. However, it is possible for us to learn how to control our own uses of technology. The 'forum' that I think is best suited for this is our educational system. If students get a sound education in the history, social effects and psychological biases of technology, they may grow to be adults who use technology rather than be used by it."[10]

In his *Teaching as a Subversive Activity*,[11] Postman states his belief that the educational system has failed many students because it does not teach or prepare them to deal with change, which is a societal constant. While this task may seem daunting, Postman believes it is possible to equip young Americans with the concepts and strategies they need to survive in a rapidly changing world. Postman joins a cadre of insightful scholars, including John Gardner, Marshall McLuhan, and Norbert Wiener, who believe it is possible for innovation to improve the human condition. What is critical now is to first figure out how to use fundamental intellectual and emotional realities to go beyond the "three Rs" and actually teach students how to learn.

As an example of how to do this, Postman and Weingartner suggest a "Questions Curriculum," of consciousness-raising exercises that reveal "new and relevant lines of inquiry," which sometimes leads to the wonderful "discovery that one's original question is far less significant than two or three others it suggests." The use of questions to construct individual meaning casts the journalistic quality of relevance in a new, more powerful and workable light.[12] In a chapter titled, "What's Worth Knowing" (they also wrote chapters on "Crap Detecting" and "Pursuing Relevance"), Postman and Weingartner offer a critical inquiry on the language of news reporting. Here is a sample of their questions:

- What is "news" anyway?
- What are its purposes?
- What is a fact?
- What do we mean by "objectivity"?
- From whose point of view is news written?
- How can you tell?
- What standards may reasonably be used to evaluate news?
- In what sense can the language of news be said to be "true"?

Lest I inaccurately represent Postman's (and also Weingartner's) ideas and thinking, I will repeat their reluctance to offer a prescriptive "how-to list," as is often the case with many experts, about how best to navigate the newsphere and the American technopoly, Postman's description of unconscious acceptance and devotion to technology. "No one is an expert on how to live a life," they say.[13]

However, they do offer some practical and useful strategies for increasing awareness, resisting dangers, and waking up to the individual responsibility now required. Postman and Weingartner recommend becoming loving resistance fighters (LRFs), people who:

- pay no attention to a poll unless they know what questions were asked and why;
- refuse to accept efficiency as the pre-eminent goal of human relations;
- freed themselves from the belief in the magical powers of numbers, do not regard calculations as an adequate substitute for judgment, or precision as a synonym for truth;
- refuse to allow psychology or any "social science" to pre-empt the language and thought of common sense;
- are suspicious of the idea of progress, and who do not confuse information with understanding;
- do not regard the aged as irrelevant;
- take seriously the meaning of family loyalty and honor, and who, when they "reach out and touch someone," expect that person to be in the same room;
- take the great narratives of religion seriously and who do not believe that science is the only system of thought capable of producing truth;
- know the difference between the sacred and the profane and who do not wink at tradition for modernity's sake;
- admire technological ingenuity but do not think it represents the highest possible form of human achievement.[14]

W. James Potter Suggests Paying Closer Attention to the Framing of News

In his *Media Literacy*, W. James Potter is more specific than Postman and Weingartner in his recommendations regarding the conscious consumption of news. Potter offers news consumers detailed strategies to deconstruct media messages: "You need to get out of the automatic processing state and devote some conscious attention to the content of certain media messages," says Potter.[15] Simple exposure to news is not enough: to wake up from that dangerous illusion, Potter makes five specific recommendations:

1. Analyze the news perspective
2. Search for context
3. Develop alternative sources of information
4. Be skeptical about public opinion
5. Expose yourself to more news, not less[16]

As we saw earlier when we looked closely at the story of Captain Richard Phillips and the ongoing reporting of piracy by Somalis, news is a story about a real event. As such, "news does not *reflect* reality," says Potter. "Instead, it is a construction by journalists. News coverage is triggered, of course, by actual occurrences. But what we see presented as news by the media are not the events themselves. Instead, the media present us with stories *about* the events, and those stories are constructed by journalists who are influenced by constraints that are largely outside of their control," says Potter.[17] He uses the term "news perspective" to cover the professional judgment, story framing, and conventional practices that constitute the construction of news.

It may appear paradoxical, but Potter recommends initially ingesting more news (not less) to learn how to develop higher level filtering capabilities. Like Postman and Weingartner, Potter outlines a variety of cognitive, emotional, aesthetic, and moral skills and knowledge needed to effectively process news and information in the digital environment. Cognitively he asks us to develop the ability to "compare and contrast key points of information in the news story with facts in your knowledge structure," an immediate way to determine its relevance and also to locate areas of ignorance.

In contrast to objective postures and framing, he recommends that newsreaders emotionally connect with the people in stories—to understand their feelings and ideally empathize with them. He recommends viewing reporting and news generation as a specific genre and comparing and contrasting "the artistry used to tell this story with that used to tell other stories." Finally, there is ethics and the requirements of a "highly developed moral code" for the producers, consumers, and distributors of journalism.[18]

John McManus Calls for Vigilance and Personal Accountability

In his interactive CD book, *Detecting Bull: How to Identify Bias and Junk Journalism in Print, Broadcast and on the Wild Web*, journalism professor John McManus offers detailed and specific strategies for testing sources, dealing with overwhelm, and recognizing the manipulation of attention. McManus brings a growing urgency to his suggestions: He believes personal responsibility and vigilance are of critical importance to conscious news consumers, who must determine for themselves whether the new forms of news stories they receive on sophisticated networks are indeed the truth. This responsibility grows daily as the fact-checking and filtering function of professional journalists declines, according to McManus.[19]

In addition to the inability to rely on professional journalists to verify the accuracy of information, other trends, such as the increasing sophistication of those who deliberately choose to manipulate information and deceive the public, are compounding this challenge and adding to its urgency: "[There] is a new sophistication in information control by people and institutions of power," Bill Kovach, Founding Chairman and Acting Director of the Committee of Concerned Journalists, told his audience when he spoke about "A New Journalism for Democracy in a New Age" in Madrid, Spain, in 2005.[20] To first recognize and then fight against this deliberate manipulation, McManus recommends that newsreaders map patterns of bias within individual outlets and across the newsphere. According to McManus, bias can be detected by asking these questions:

1. Who is the author, who is the sponsor, and what conflicts exist between their self-interest and the public interest?
2. What is the purpose of the article—to inform, to capture our attention for advertisers, to sell us goods or services, or to package a press release as news?
3. Which groups affected by the issue or event reported are represented among the sources quoted, and which are missing?
4. What evidence is offered for fact-claims?
5. Who gains and who loses from the way the story is framed [using omissions, value words, stereotypes, metaphors, themes, and story structures]?
6. How do other media report the same issue or event?

In addition to the bias in outlets and individual stories, McManus has constructed an "image bias detector," a critical lens to view images and video clips. He suggests looking carefully at the stakeholder group, the foreground and background, and the key characters; and to carefully consider whether the players are active (aggressor) or passive (victim). For example, who is sympathetic and who is neutral? What is the primary emotion elicited? He recommends carefully considering who is not pictured in photographs and video clips and also thinking about whether the image is consistent with its accompanying article or story. The power of manipulated image is hard to estimate, as McManus brilliantly illustrates when he replaces the faces of a pair of professional ice skaters in matching outfits with the faces of Barack Obama and Sarah Palin, a most unlikely duo.

Howard Rheingold's Antidote for News Pollution

There is an antidote for information "tainted by ignorance, inept communication, or deliberate deception," according to Howard Rheingold, whose long history of digital literacy began with the WELL in San Francisco, one of the Internet's first virtual communities. "I believe that what people know—and how many people know—matters. Digital media and networked publics are only the infrastructure for participation—the cables and chips do no good unless people know how to use them," Rheingold writes in his "Crap Detection 101," a staple on the news literacy site, NewsTrust.net, which is widely circulated on the Web. According to Rheingold, "the good stuff is out there if you know how to find it."[21]

The following categories of questions often a powerful lens to "find the good stuff," as Rheingold recommends, and eliminate "the crap."

Currency
> How recent is the information?
> How recently has the website been updated?
> Is it current enough for your topic?

Reliability
> What kind of information is included in the resource?
> Is the content of the resource primarily opinion? Is it balanced?
> Does the creator provide references or sources for data or quotations?

Authority
> Who is the creator or author?
> What are the credentials?
> Who is the publisher or sponsor?
> Are they reputable?
> What is the publisher's interest (if any) in this information?
> Are there advertisements on the website?

Purpose and Point of View
> Is this fact or opinion?
> Is it biased?
> Is the creator or author trying to sell you something?

Some Simple Rules for Navigating the Newsphere

In addition to the advice of these experienced and knowledgeable experts, try these simple rules for navigating the newsphere, finding a comfortable place in it, and designing an individual filter and news network:

1. Limit your exposure to negative news, people, and events.
2. Use your body to pay closer attention to how news stories affect you personally. Pay close attention to how you feel when you take in and process news. Don't let it simply wash over you. Be alert to what you think and especially what you feel when you hear or learn something new. That is a clue to its relevance, value, and importance to you and your life. Your body is your primary medium of communication. It has much wisdom to share with you if you take good care of it and quietly listen for its messages.
3. Observe instead of judge. Build in reaction time. Create a space between the unfolding of an event and the intake of news and information, and

give yourself time for reflection before you act. Watch what you think because your thoughts will determine your reality.

4. Choose instead of trying to change.
5. Recognize when you are in balance and stay there as much as possible.

How the *Ann Arbor Chronicle* Creates Value

By Helen Nevius Adamopoulos

Dave Askins, editor and cofounder of the *Ann Arbor Chronicle*, sums up the work his publication does in two words: not sexy. While other reporters set out looking for drama and flaring tempers, the *Chronicle* concentrates on what many would consider dull: downtown development authority meetings, library board work sessions, and city council decisions. Covering local government might not be exciting, but Askins and his wife, co-editor and the *Chronicle's* publisher Mary Morgan, agree that it's necessary. "I consider it to be one of those fundamental axioms that's obvious," Askins said. "It's intrinsically important."

Since the *Chronicle's* premiere in September 2008, Askins and Morgan have supplied the residents of Ann Arbor, Michigan, with comprehensive coverage of local legislative bodies, such as the Ann Arbor Public Art Commission and the city's planning commission. They approach local government using a different strategy than most journalists: Instead of choosing one issue to write about, they report everything that happens at meetings. As the publication's tagline states, "it's like being there." Askins said this approach allows readers to follow certain issues and review decisions. He said it's also helpful to have an alternative to the official meeting minutes, which aren't legally required to provide any detail beyond who voted for what. He added that the publication is not only informing present-day Ann Arbor residents but is also recording crucial information for future historians. "We're writing a bit of the history of this city, allowing people to understand in a very deep and detailed way what the hell happened during our publication," Askins said.

Morgan—who spent 12 years working for the now-defunct *Ann Arbor News* before launching the *Chronicle*—describes the traditional reporting strategy concerning government meetings as the "parachute journalism approach." The reporters jump in, find one thing that seems important, and abandon the rest. "I think it's a real disservice," Morgan said. "You're filtering based on your own judgment of what's important." In addition to using a full disclosure strategy when it comes to meetings, Askins and Morgan clarify issues for their readers by translating government shorthand. Morgan said officials often utter sentences that allude to years' worth of history or refer to an obscure piece of legislation. The *Chronicle* editors will research the issue's historical background or find the relevant legislation and publish it in their meeting report in order to make things clear for their readers. Morgan said the *Chronicle* could serve as a useful resource for those who want to run for public office or just want to participate in local government in some way.

"If you read what we report, you have a very deep understanding of how government works," Morgan said. Askins notes another advantage of the *Chronicle's* reporting style: Keeping an extremely detailed record of every meeting means that public officials are more accountable. He said Ann Arbor Mayor John Hieftje has a tendency to quibble with the *Chronicle*, trying to deny some of the statements Askins and Morgan record. Other public officials also don't like the *Chronicle* for publishing everything they say. "It's much harder for elected officials to weasel out of positions," Morgan said. The *Chronicle* publishes 10 to 14 articles a week. In addition to its founders, the news outlet has about a dozen regular contributors and provides limited coverage of sports and the arts. Askins and Morgan support themselves and the publication with a combination of advertising revenue and subscriptions. The *Chronicle* editors work every day, for most of their waking hours.

Askins and Morgan frequently have people thank them or praise them for what they do, whether it's by approaching them in person or by leaving a comment on the *Chronicle's* website. Askins recalls a local restaurant owner using one of their reports on the city's historic district as guidance for a related project his business was about to undertake. Morgan said she heard that a teacher passed a *Chronicle* article around the classroom because of its detailed description of how a millage works. "Somebody said he's lived in this town for 15 to 20 years, and it's only been since we started publishing that he understands what's going on," Morgan said. She notes that there's a general sentiment that nobody cares about local government. However, the *Chronicle's* 30,000-plus visitors per month prove that wrong. "We don't need every single person in the community to read us, and I don't think every single person is interested in what we do," Morgan said. "But enough are."

This chapter offered a wide range of strategies including the comprehensive reporting style of the *Ann Arbor Chronicle* as suggestions for learning how to consciously navigate the newsphere. As McManus and Rheingold note, it is a time of urgency and vigilance requiring reflection and careful consideration of Postman's question—"What is news, anyway?" The next and final chapter, *The Essence of Journalism*, explores the new role of news consumers and journalists, and carefully considers both the essence of journalism and more evolved qualities of news.

ENDNOTES

1 I conducted several interviews with Sarah Merion via email and one Skype interview from Buenos Aires during the spring of 2010.

2 Postman, N. (1992). *Technopoly: The Surrender of Culture to Technology.* New York: Vantage Books. p. 60.

3 Ben Elowitz (@elowitz) is cofounder and CEO of Wetpaint, a platform for social websites, and author of the Digital Quarters blog. These remarks are from Elowitz's "Leading Voices blog, May 3, 2010" and "If the News Is That Important, It Will Find Me" in "Finding Political News Online: The Young Pass It on" by Brian Stelter, March 27, 2008. Retrieved February 14, 2011. http://www.nytimes.com/2008/03/27/us/politics/27voters.html. Elowitz's work can also be found at http://paidcontent.org/topic/leading-voices/.

4 Retrieved February 15, 2011. http://www.guardian.co.uk/media/2009/jun/26/michael-jackson-tmz-scoop. *The New York Times* reported on this phenomenon in a Sunday, May 22, 2011, front page story titled, "The Gossip Machine, Churning Out Cash."

5 http://twitpic.com/135xa. Retrieved November 8, 2010.

6 Reported on the "Cult of Mac" blog. Available at http://www.cultofmac.com/space-shuttle-endeavor-launch-also-captured-on-video-by-iphone/95608.

7 p. 2 of 16 on February 22, 2010. www.buzzmachine.com. Retrieved February 23, 2010.

8 Teilhard, P. (1966). Demoulin, J. (Ed.). *Let Me Explain*. New York: Harper & Row. p. 24.

9 Teilhard, P. (1966). Demoulin, J. (Ed.). *Let Me Explain*. New York: Harper & Row. p. 67.

10 *PBS NewsHour* interview: "Neil Postman Ponders High Tech." (1996, January 17). Retrieved July 29, 2010 from http://www.pbs.org/newshour/forum/january96/postman_1-17.html.

11 Postman, N. *Teaching as a Subversive Activity*. p. xiv.

12 Postman, N. *Teaching as a Subversive Activity*. p. 70.

13 Postman, N. *Technopoly*. p. 182.

14 Postman, N. *Technopoly*. pp. 183–185.

15 Potter, W. *Media Literacy*. p. 187.

16 Potter, W. *Media Literacy*. pp. 187–188.

17 Potter, W. *Media Literacy*. p. 171.

18 Potter, W. *Media Literacy*. p. 189.

19 McManus. *Detecting Bull: How to Identify Bias and Junk Journalism in Print, Broadcast and on the Wild Web*. p. 4.

20 Kovach, B. (2005, February 1). "A New Journalism for Democracy in a New Age," speech given in Madrid, Spain, and available at http://www.journalism.org/.

21 http://www.newstrust.net/guides/crap-detection-101.

22 http://www.workliteracy.com/the-crap-test.

· 7 ·

THE ESSENCE
OF JOURNALISM

*We are one, after all, you and I. Together we suffer, together exist,
and forever will re-create each other.*
                                        —*Pierre Teilhard de Chardin*

The Origins of the Word *Noosphere*

Pierre Teilhard de Chardin drew profound inspiration from his service as a chaplain and noncombatant stretcher-bearer during World War I, a tragic and horrific chapter in human history. Teilhard began his assignment with the North African Zouaves in January of 1915 after learning that his younger brother, Gonzague, had been killed in battle near Soissons in northern France. During his years of service, and indeed as was a custom throughout his life, Teilhard kept in regular correspondence with his family and friends. His letters to his cousin, Marguerite, are now collected in *The Making of a Mind*, and they offer a glimpse of the power, grace, and compassion of Teilhard the man.[1]

In his writings to Marguerite he poetically describes witnessing the horrors and heroism of war—"things that you experience nowhere else." The tension, courage, and pain of a solder's day-to-day life on the battlefield opened his heart and his mind and pushed him to see more deeply into the essence of reality. In the trenches of World War I, he was able to recognize the existence of "the extreme boundary between what one is already aware of, and what is still in process of formation." This awareness forever changed him. He discovered an "underlying stream of clarity, energy, and freedom" that he carried into the rest of his life. On the battlefield, he witnessed the unity of all souls and was able to fully and consciously embrace the "quasi-collective life of all men."

Teilhard invented the word *noosphere* to describe and represent humanity's collective consciousness. As he envisioned it, the noosphere is the soul of the earth, and as such, it "represents the total pattern of thinking organisms and their activity, including the patterns of their interrelations." Its power and potential goes beyond serving as a simple container or environment in the biosphere above the earth; the noosphere is both a repository as well as the "impetus for the growth of intelligence through human thought and action."[2]

Journalists tap and foster the power of this intelligence when they build relevant and meaningful associations in and with their news stories. Together journalists and the public weave a dynamic real-time news fabric that encourages readers to follow a story and determine its personal relevance while reporters continually monitor its veracity, progress, and evolution. This style of news creation also uncovers often hidden assumptions so they can be re-evaluated. It places the individual at the center and assumes cooperation and participation. In a safe collaborative space, informed individuals thus have the support to discover the truth.

News looks different through a Teilhardian lens. *When viewed as an ecology, news is not a product to be consumed, but a conscious act of engaging with and producing shared information that has value in a community. This is how cultures and societies create their histories.* News is a reflection, not a construction of reality. It is information that individually and collectively adds to public knowledge, promotes global understanding, and encourages engagement. Above all, it is not divisive. Instead, integral journalism's dialogue-style of news unites. It creates new connections and reinforces old ones through a nonjudgmental mixture of listening and speaking. In this way, meaning is co-created by the collective.

With journalists as conversational partners, the public can more readily and easily engage in relevant and meaningful discussions, which are more possible now because of the accessibility and efficiency of two-way digital communication tools. Dialogue-style news is ideally suited to supporting a journalism of conversation because it enables many levels of collaboration. When dialogue and not debate is the focus of news coverage, then the public can more readily see all sides of an issue and re-evaluate their assumptions, as they search for the strength and values in others' positions instead of simply defending their existing beliefs and convictions.

New Qualities of News

In its most evolved form, news in the newsphere exhibits the following qualities: It is *networked, participatory, relevant, and transparent.* News is *networked* now because evolutionary forces of change keep the newsphere open and fundamentally self-vetting. Networked, open, transparent content vets itself. When it is freely accessible, news can connect, patterns can arise, and meaning is acquired. This is the new value and the new sustainability. When connections enhance a free flow of information, accuracy follows in a timely way. An open news environment is structured to produce the truth, something we can all claim as our collective intelligence. A news story gains value as it evolves through the intelligent networks of the heterarchical newsphere. It contains information that individually and collectively adds to public knowledge, promotes global understanding, and encourages engagement. Above all, it is not divisive. Instead, a dialogue-style of news unites. It creates new connections and reinforces old ones through a nonjudgmental mixture of listening and speaking.

A connective, integral style of journalism fosters a high level of *participation* and engagement. Inspiring involvement with the newsphere is especially important now because many people have simply stopped receiving news because it is so negative, detached, and official. By revaluating what constitutes news and embracing these new qualities, such as networked and participatory, journalists are better able to promote the meaningful dialogue, conversation, and community building necessary to not only engage their audiences but to also support political action and more evolved forms of government.

French scholar Pierre Lévy's real-time democracy is an illustrative example of an evolved and emerging political institution. Writing in his *Collective Intelligence,* Lévy describes a real-time democracy as a dynamic institution that helps the public share news and information more efficiently and effectively. It is founded on the self-determination of an informed and enlightened public: "Real-time democracy maximizes the responsibility of the citizen called upon to make decisions, accept the consequences of those decisions, and make judgments on the basis of those decisions,"[3] says Lévy. He believes that a commitment to collective intelligence will lead to the creation of the technological tools, means of communication, and new social structures and organizations that will "enable us to think as a group, concentrate our intellectual and spiritual forces, and negotiate practical real-time solutions to the complex problems we must inevitably confront."[4]

The exercise of personal responsibility is the driving force behind the creation and sustainability of the enlightened and collective consciousness necessary for the formation of Lévy's "real-time democracy" and Teilhard's "le Tout," a unified reality. The collective has a moral imperative to shape and refine the environment. This means navigating the newsphere consciously. While a news story can be fair and accurate, real truth evolves over time, based in large part on the demands placed by the people who ideally crave, consume, and use it. The value which is brought to journalistic fare, an individual and collective demand for *relevance*, will ultimately result in the quality so desperately needed now.

So how do enlightened journalists wade into the vast sea of current events and determine what is indeed newsworthy and *relevant* to audiences? Matt Thompson, a Reynolds Journalism Institute Fellow at the University of Missouri and former deputy web editor of the *Minneapolis Star Tribune*, posed the following questions on his Newsless.org blog as a guide for journalists:

1. Are we making our community feel better informed or merely distracted?
2. How important is this for our community to know and why?
3. Are we chasing the larger story, or just the latest story?
4. Are we synthesizing information, or merely aggregating it?
5. How are we serving those who know {nothing | a lot} about the topic?
6. Have we provided a clear trail through our coverage?
7. Are we using 1,000 words where a picture should be?
8. How good are our filters?
9. Will our coverage find its audience where and when they're ready for it?
10. How are we managing our own information overload?

New Forms of News

New genres of news, such as blogs, games, feeds, and forums, as well as more traditional practices, such as open-source reporting and contributor profiles, allow for an unprecedented level of *transparency*, the final of the four qualities required in the newsphere. The interactive and accessible design of new forms of news, particularly those that support community-building through regular engagement, encourage honesty and offer pioneering opportunities for building credibility. Likewise, open-source reporting, while not a new practice, fosters transparency. It asks readers to join with professional reporters in the news creation process through the submission of leads, sources, tips, and topics worthy of investigation.

Historically, news organizations have offered "tip lines." However, the ease and accessibility of news platforms and tools keep expanding these opportunities for citizen journalists. In addition to open-source reporting and the ease of access to contributor profiles, other innovative practices, such as distributed reporting, allow readers to submit actual news reports that are then collated in a database and widely disseminated. These practices will continue to prosper as networks expand, technological tools become more accessible, and audiences embrace their emerging power as both consumers and producers of news.

Innovative and experimental forms of news, such as such as interactive news-games, support the co-creation that is critical for the energetic union that Pierre Teilhard de Chardin believed would heal differences and promote planetary evolution. There are many excellent examples of innovative news practices and quality journalism in the current newsphere, such as Ken Burn's PBS series *Prohibition*, as well as in a long history of American literary journalism. Researchers at Georgia Tech are pioneering newsgames as an engaging and aesthetically pleasing new form of news that invites users to participate in real-world simulations.

The Georgia Tech researchers believe their games "persuade, inform, and titillate; make information interactive; re-create a historical event; put news content into a puzzle; teach journalism; and build a community."[5] One example of their work is "Sweatshop," which invites players to assume the role of a factory floor manager in a developing nation. Players are challenged to move through 30 different stages, placing skilled workers and children along a conveyor belt. They are then scored on the efficiency and the character of their management decisions in this enactment of the high human cost of sweatshops producing low-cost consumer goods.

The New Role of News Consumers

News is no longer a spectator sport. Integral journalism places the individual—not the news story or the event—at the organic center of the news-generation process. This new style of news demands a heightened sense of awareness and perception—a genuine openness to the truth and the reality. It harnesses the power and inherent responsibility of each citizen and uses it for the collective good. It is, above all, connective. Embracing these new duties means: looking for original reporting in all its forms and guises; avoiding "churnalism," those stories that merely rehash and spin existing information; and recognizing "news about news" and debate-style stories. This remains a daunting task because whistle-blowers and truth tellers are often punished in the newsphere, while those who provide the comfort of illusion are unfortunately rewarded.

As the concept of "backfire" illustrated, individual belief systems and perceptions—how and what we see—dramatically alter the interpretation, understanding, and integration of news stories. To see as Teilhard saw requires an upgrading of perceptive abilities. Newsreaders and users must now understand that the truth—a personal and a collective truth—is embedded somewhere in a news story and now they must dig for it! They will need sophisticated emotional, aesthetic, and moral skills and knowledge to combat the news creators and disseminators who want to manipulate their time, attention, and emotions.

It is time to consider becoming one of Postman and Weingartner's LRFs (loving resistance fighters). LRFs ask and ideally answer questions such as, "What is news, anyway?" and "How can you tell?" This critical process builds news literacy skills and an organic understanding that truth comes from the inside out. Digging for the truth is especially challenging in the current news environment, where singular opinions are elevated to the status of tangible and verifiable truth. A powerful example of this phenomenon is *The Rush Limbaugh Show*, the highest-rated talk-radio program in the United States.[6] Limbaugh, who attended two semesters of college before dropping out, is paid $50 million each year to share his beliefs, views, judgments, and attitudes on the Clear Channel network.[7]

As Neil Postman, the founder of the media ecology tradition, has pointed out, "*an opinion is not a momentary thing but a process of thinking shaped by the conscious acquisition of knowledge and the activity of questioning, discussion, and debate.*" Opinions must be examined and adjusted through the organic vetting processes of the newsphere. Opinions such as those disseminated so widely and so profitably by Limbaugh and all commentators must be very closely examined and questioned by audiences. As Ernest Hemingway wrote in 1954 and media literacy scholar Howard Rheingold reminds readers on his "Crap Detection 101" site, "Every man should have a built-in automatic crap detector operating inside him."[8]

Often new media technologies have exacerbated the elevation of opinion to truth. Instead of adding new information to support the veracity of a story or enhancing enterprise reporting, ever-evolving technological tools are often used only as a distribution channel, or worse. Tempted by large advertising profits, employees of the United Kingdom's *News of the World* tabloid newspaper used sophisticated devices to routinely hack into cell phones. Their list of victims includes celebrities, politicians, and Britain's Royal Family, as well as murdered schoolgirl Milly Dowler, relatives of deceased British soldiers, and victims of the July 7, 2005, London bombings. The resulting "Murdochgate" scandal closed the newspaper and raised difficult and complex questions about press regulation, media ownership, and the relationships between politicians and journalists.[9]

Listening Rhetoric Informs News Consumption

Developing an ideology of news and a practical understanding of the newsphere is an exciting, important, and challenging task. In addition to media ecology, journalism, and critical media studies literature, this analysis also relies on rhetorical theories and theorists. Rhetorical theory offers a powerful lens for viewing the emergence of new journalistic forms and practices and also for critical evaluation of its function in a democratic society. As we saw in Chapter Two, "The Illusion of News," the rhetorical challenges facing both the public and practicing journalists are immense. Insightful rhetorical scholarship is needed to clarify and elucidate how news is changing, the myth of objectivity, the role and function of the citizen journalism movement, the institutionalizing of professional practices, and numerous other embedded concepts and practices.

Two specific definitions of rhetoric were used here. University of Colorado Professor Gerard A. Hauser defines rhetoric as "the moment-to-moment construction of reality."[10] Walter Ong defines rhetoric as "the mindful negotiation of the realm between the conscious and the unconscious."[11] A synthesis of Hauser and Ong's definitions implies a resolution or unity. Thus rhetoric, as it is applied in this analysis, is seen as an active effort to unify the conscious and the unconscious on both an individual and a societal level. This rhetorical unity is evidenced in the work of Wayne Booth.

In his *Rhetoric of Rhetoric*, Booth describes and celebrates the dynamic unity possible when someone speaks the truth and someone else really listens with an open mind and heart. He calls it Listening Rhetoric (LR): "Here both sides join in a trusting dispute, determined to listen to the opponent's arguments while persuading the opponent to listen in exchange. Each side attempts to think about the arguments presented by the other side. Neither side surrenders merely to be tactful or friendly.…Both sides are pursuing not just victory but a new reality, a new agreement about what is real."[12]

Unlike the debate style of news that dominates political discourse and other kinds of rhetoric, such as the "win rhetoric" that is too familiar and too often practiced, Booth wisely notes that the only consequence of too much listening rhetoric is "self-censorship." He advocates that the "training of everyone to pursue critically the defensible kinds of rhetoric is one of our best hopes for saving the world—or at least this or that corner of it."[13]

The New Role of the Integral Journalist

Integral journalism places the individual at the center and assumes that collaboration and participation are necessary. It supports listening not only to understand, but also as a means to find meaning and agreement, and uncovers often hidden assumptions so they can be re-evaluated. It helps explore common ground and supports dialogue. In a safe collaborative space, informed individuals have the support to discover the truth. This collaborative and open style of news makes it possible to move beyond entrenched ways of thinking to explore and perhaps embrace new ideas and information. Above all, integral journalism is dynamic and open-ended. Topics are kept alive as citizen journalists search for the strength and value in others' positions. Likewise, integral journalists are able to exercise their new judgment in powerful new ways. This has become increasingly important in the newsphere, where many people have simply stopped receiving news because it makes them feel fearful, isolated, and powerless.

The challenge today is to bring the energy and insight Teilhard displayed into the newsphere—our global news and information network. Viewing news through a Teilhardian lens means perceiving and believing in humanity's organic connectedness and acting from this place. It means using ever more sophisticated digital tools to build links and keep news stories open and self-vetting. That is the production part—the work of the journalist. It also requires action and courage to first recognize and then challenge the ignorance, lies, and propaganda that veil the truth and keep us divided. That is the consumption part—everyone's job.

The rigidly opinionated must not continue to control the newsphere and boisterously lead the armies of manipulation while the public stands idly by, or worse, simply abandons the fight. The newsphere is sustained by the energy of integral journalists committed to a dialogue style of news. It is composed of news consumers who keep their hearts and minds open and listen with the possibility that they might change their thinking about long-held beliefs and conviction. It is fueled by intelligent networks that work individually and collectively to keep content open and self-vetting.

Above all, it gives witness and is committed to the truth.

ENDNOTES

1　Taken from the bibliography of the American Teilhard Association, which is available at http://teilharddechardin.org/index.php/biography. *The Making of a Mind.* p. 205.

2　Teilhard, P. (1966). p. 7.

3　Lévy, P. *Collective Intelligence.* p. 73.

4　Lévy, P. *Collective Intelligence.* p. xxvii.

5　See http://jag.lcc.gatech.edu/blog/. Retrieved August 18, 2011.

6　See http://en.wikipedia.org/wiki/Rush_Limbaugh.

7　Stelter, B. (2008, July 3). A lucrative deal for Rush Limbaugh. *The New York Times.* Edition retrieved September 8, 2010.

8　See http://www.sfgate.com/cgi-bin/blogs/rheingold/detail?entry_id=42805.

9　See http://en.wikipedia.org/wiki/News_International_phone_hacking_scandal.

10　Hauser presented this definition at the 2002 biannual meeting of the Rhetoric Society of America, held in Las Vegas, NV.

11　Ong, W. *Rhetoric, Romance, and Technology.* pp. 11–12.

12　Booth, W. (2004). *The Rhetoric of Rhetoric.* Malden, MA: Blackwell Publishing. p. 47.

13　Booth, W. p. 54.

BIBLIOGRAPHY

Aczel, Amir D. (2007). *The Jesuit and the Skull*. New York: Riverhead Books.

Aiello, Mary (2008). *Investigating the Blackout of August 14, 2003: Voices Left Behind in the Darkness.* Unpublished master's thesis, Eastern Michigan University, Ypsilanti.

Barnhurst, Kevin G., and Nerone, John (2001). *The Form of News*. New York: The Guilford Press.

Birx, H. James (1972). *Pierre Teilhard de Chardin's Philosophy of Education*. Springfield, IL: Charles C. Thomas Publishers.

Boczkowski, Pablo J. (2004). *Digitizing the News: Innovation in Online Newspapers*. Cambridge: The MIT Press.

Bolter, Jay David, and Grusin, Richard (1999). *Remediation: Understanding New Media*. Cambridge: The MIT Press.

Boyman, Shayne, and Willis, Chris (2005 Winter). The future is here, but do news media companies see it? *Nieman Reports: Citizen Journalism 59(4)*. Available at www.nieman.harvard.edu.

Bridle, Susan (n.d.). The divinization of the cosmos: an interview with Brian Swimme on Pierre Teilhard de Chardin. Retrieved from http://www.enlightennext.org/magazine/j19/teilhard.asp?pf=1. Downloaded January 18, 2011.

Burrough, Bryan, Peretz, Evgenia, Rose, David, and Wise, David (2004, May). The path to war. *Vanity Fair (525)*, 228-294.

Bye, Kent (2006, May 10). Building a theory of collaborative sensemaking. *The Echo Chamber Project*. Available at http://www.echochamberproject.com/about.

Bye, Kent (2005, January 11). Integral journalism—a new paradigm [Web log post]. Retrieved from http://echochamberproject.com/node/62. Downloaded November 18, 2010.

Carey, James W. (1989). *Communication as Culture: Essays on Media and Society*. Winchester, MA: Unwin Hyman.

Carey, James W. (1993 Summer). The mass media and democracy: between the modern and the postmodern (Power of the Media in the Global System). *Journal of International Affairs 47(1)*,

1-21. *InfoTrac OneFile.* Thomson Gale, Eastern Michigan University, 8 Nov. 2006. Retrieved from http://find.galegroup.com/itx/infomark.do?&contentSet=iac-Document&type=retri eve&tabID=T002&prodid=ITOF&docId=A14469040&source=gale&srcprod=ITOF&. Downloaded November 10, 2007.

Carey, James W. (1996 January/February). The struggle against forgetting. *Columbia Journalism Review.* Available at http://www.jrn.colombia.edu/admissions/struggle/.

Cathcart, Robert S. (1993). Instruments of his own making: Burke and media. In James W. Chesebro (Ed.), *Extensions of the Burkeian System* (pp. 287–308). Tuscaloosa: The University of Alabama Press.

Crick, Nathan (2009). The search for a purveyor of news: The Dewey/Lippmann debate in an internet age. *Critical Studies in Media Communication, 23*(5), 480–497.

Cunningham, Brent (2003, July 9). Rethinking objective journalism. *Columbia Journalism Review.* Retrieved from http://alternet.org/media/16348?page=entire. Downloaded July 26, 2010.

Darnton, Robert (1975). Writing news and telling stories. *Daedalus: The Journal of American Academy Arts and Sciences, 104,* 173–195.

Demoulin, Jean-Pierre (1966). *The Essential Teilhard: Let Me Explain Pierre Teilhard de Chardin.* New York: Harper & Row.

Dery, Mark (2004, October 10). Culture jamming: hacking, slashing and sniping in the empire of signs. [Web log article]. Retrieved from http://www.markdery.com/archives/books/ culture_jamming/#0....

deWit, J.J. Deyvene (1996). "Pierre Teilhard de Chardin" In Philip Edgcumbe Hughes (Ed.), *Creative Minds in Contemporary Theology.* Grand Rapids, MI: Wm. B. Eerdmans Publishing Co.

Eisenstein, Elizabeth L. (1979). *The Printing Press as an Agent of Change: Communications and Cultural Transformations in Early-Modern Europe 1.* Cambridge, UK: Cambridge University Press.

Eisenstein, Elizabeth L. (1983). *The Printing Revolution in Early Modern Europe.* Cambridge, UK: Cambridge University Press.

Farrell, Thomas J. (2000). *Walter Ong's Contribution to Cultural Studies: The Phenomenology of the Word and I-Thou Communication.* Cresskill, NJ: Hampton Press.

Farrell, Thomas J. (2003). "Walter Ong's thought as framework and orientation for cultural studies in the humanities." *Renascence,* Summer 2003. Retrieved from http://www.findarticles. com/p/articles/mi_qa3777/is_200307/ai_n928. Downloaded October 15, 2008.

Fidler, Roger (1997). *Mediamorphosis: Understanding New Media.* Thousand Oaks, CA: Pine Forge Press.

Gordon, Eric, and Bogen, David (2009 Spring). Designing choreographies for the "new economy of attention." *DHQ: Digital Humanities Quarterly, 3*(2). Retrieved from http://digitalhuman- ities.org/dhq/vol/3/2/000049/000049.html.

Greene, Brian (2004). *The Fabric of the Cosmos: Space, Time, and the Texture of Reality.* New York: Vintage Books.

Grim, John, and Tucker, Mary Evelyn (2004–2005). Teilhard de Chardin: a short biography. Retrieved from http://teilharddechardin.org/biography.html. Downloaded February 3, 2010.

Gumpert, Gary (1985). Media grammars, generations, and media gaps. *Critical Studies in Mass Communication 2*, 23–35.

Gumpert, Gary and Cathcart, Robert (Eds.). (1986). *Inter Media: Interpersonal Communication in a Media World.* (3rd ed.) New York: Oxford University Press.

Habermas, Jürgen (1964). The public sphere: an encyclopedia article. (Sara Lennox and Frank Lennox, Trans.) Originally appeared in Fischer Lexicon, *Staat und Politik* (Rev. ed.), (pp. 220-226). Frankfurt am Main.

Harty, Sheila T. (1999). *Teilhard de Chardin's Theology of Evolution.* Retrieved from http://ebooks-browse.com/teilhard-de-chardins-theology-of-pdf-d185227765.

Herman, Edward S. and Chomsky, Noam (1988). *Manufacturing Consent: The Political Economy of the Mass Media.* New York: Pantheon Books.

Hiler, John (2002, May 28). Blogosphere: the emerging media ecosystem. *Microcontent News.* Available at http://www.microcontentnews.com/articles/blogosphere.htm.

Innis, Harold A. (1972). *Empire and Communications.* Toronto: University of Toronto Press.

Innis, Harold A. (1991). *The Bias of Communication.* Toronto: University of Toronto Press.

Jankowski, Nicholas W., and Van Selm, Martine (2001). Traditional news media online: an examination of added values. In K. Renckstorf, D. McQuail, and N. Jankowski (Eds.), *Television News Research: Recent European Approaches and Findings* (pp. 375–392). Berlin: Quintessence Publishing Co.

Jarvis, Jeff (2010). BuzzMachine [Web log post]. Retrieved from http://buzzmachine.com/...

Kamen, Al (2007, October 26). FEMA meets the press, which happens to be…FEMA. *The Washington Post.* Retrieved from http://www.washingtonpost.com/wp-dyn/content/article/2007/10/2.... Downloaded November 10, 2007.

Kaufer, David S., and Carley, Kathleen (1993). *Communication at a Distance: The Influence of Print on Sociocultural Organization and Change.* Hillsdale, NJ: Lawrence Erlbaum Associates Publishers.

Kawamoto, Kevin (Ed.) (2003). *Digital Journalism: Emerging Media and the Changing Horizons of Journalism.* Lanham, MD: Rowman & Littlefield Publishers.

Keating, Maurice, and Keating, H. R. F. (1996). *Understanding Pierre Teilhard de Chardin.* London: Lutterworth Press.

Keohane, Joe (2010, July 11). How facts backfire: workers discover a surprising threat to democracy: our brains. *The Boston Globe.* Retrieved from http://www/boston.com/bostonglobe/ideas/articles/2010/7/1...

King, Ursula. (1996). *Spirit of Fire: The Life and Vision of Teilhard de Chardin.* Maryknoll, NY: Orbis Books.

Kolbert, Elizabeth (2003, June 30). Tumult in the classroom. *The New Yorker.* Retrieved from http://www.newyorker.com/archive/2003/06/30/03063fa_fact?.... Downloaded July 19, 2010.

Konner, Joan (1996, January/February). The struggle against forgetting. *Columbia Journalism Review, 1.* Retrieved from http://backissues.cjrarchives.org/year/96/1/pubnote.asp. Downloaded August 16, 2007.

Kostelnick, Charles, and Hassett, Michael (2003). *Shaping Information: The Rhetoric of Visual Conventions.* Carbondale: Southern Illinois University Press.

Kuhn, Thomas S. (1970). *The Structure of Scientific Revolutions* (2nd ed.). Chicago: The University of Chicago Press.

Lapham, Christine M. (Tracy) (1995). "The Evolution of the Newspaper of the Future" In *The Proficient Reader,* edited by Ira Epstein and Ernie Nieratka, Houghton Mifflin, Boston, MA.1998, and in *Perspectives: Online Journalism,* edited by Kathleen Wickman, Coursewise Publishing, Inc., Bolder, CO

Lefevre, Karen Burke (1987). *Invention as a Social Act.* Carbondale: Southern Illinois University Press.

Lemann, Nicolas (2005, February 14 and 21). Fear and favor. *The New Yorker,* 168–176.

Lum, Casey Man Kong (2006). *Perspectives on Culture, Technology, and Communication: The Media Ecology Tradition.* Cresskill, NJ: Hampton Press.

Manoff, Robert Karl, and Schudson, Michael (Eds.) (1987). *Reading the News: A Pantheon Guide to Popular Culture.* New York: Pantheon Books.

McChesney, Robert W. (1999). *Rich Media, Poor Democracy: Communication Politics in Dubious Times.* New York: The New Press.

McChesney, Robert W., and Nichols, John (2002). *Our Media Not Theirs: The Democratic Struggle Against Corporate Media.* New York: Seven Stories Press.

McIntosh, Steve (2007). *Integral Consciousness and the Future of Evolution: How the Integral Worldview Is Transforming Politics, Culture and Spirituality.* St. Paul, MN: Paragon House.

McLuhan, Marshall. (1994). *Understanding Media: The Extensions of Man.* Cambridge, MA: MIT Press.

McLuhan, Marshall (1962). *Gutenberg Galaxy: The Making of Typographic Man.* Toronto: The University of Toronto Press.

McManus, John H. (2009). *Detecting Bull: How to Identify Bias and Junk Journalism in Print, Broadcast and on the Wild Web.* CD-ROM. Sunnyvale, CA: The Unvarnished Press. ISBN: 978-0-9840785-0-9.

Meyrowitz, Joshua (1998). Multiple media literacies. *Journal of Communication,* 96–107.

Min, Jeak K. (2005, Winter). Journalism as a conversation. *Nieman Reports: Citizen Journalism 59(4).* Available at www.nieman.harvard.edu.

Mott, Frank Luther (1941). *American Journalism: A History of Newspapers in the United States Through 250 Years, 1690 to 1940.* New York: Macmillan.

Noel, Hans (2010). Ten things political scientists know that you don't. *The Forum,* 8(3). Article 12. Retrieved from http://www.bepress.com/forum/vol8/iss3/art12.

Norman, Donald (1999, May). Affordance, conventions and design. *Interactions,* 38–43.

Nyhan, Brendan (2010). BrendanNyhan [Web log] http://www.brendan-nyhan.com/

Nyhan, Brendan, and Reifler, Jason (2010, March 30). When corrections fail: the persistence of political misperceptions. *Journal of Political Behavior* (2010) *32,* 303–330. doi:10.1007/s1109-010-9112-2.

Ølgod, Viki (2011). "Mapping citizen journalism in the blogosphere." Retrieved www.uta.fi/laitokset/tacs/papers0506/olgod.pdf at the Media Lab, University of Art and Design Helsinki.

Ong, Walter J. (1957). "Technology and new humanist frontiers," In *Frontiers in American Catholicism: Essays on Ideology and Culture.* New York: Macmillan.

Ong, Walter J. (1971). *Rhetoric, Romance, and Technology: Studies in the Interaction of Expression and Culture.* Ithaca, NY: Cornell University Press.

Ong, W. J. (1976). "Communications as a field of study." In S. Bamberger (Ed.), *The 1977 Multimedia international yearbook* (pp. 7–25). Rome: Multimedia International.

Ong, Walter J. (1977). *Interfaces of the Word: Studies in the Evolution of Consciousness and Culture.* Ithaca, NY: Cornell University Press.

Ong, Walter J. (1999). *Orality & Literacy: The Technologizing of the Word.* London: Routledge.

Phelan (Waite), Catherine Kaha (2002). The role of the self in the social world. *Studies in Symbiotic Interaction 24,* 215-231.

Pickard, Victor, Josh Stearns and Craig Aaron (n.d.). Saving the news: toward a national journalism strategy. Retrieved from www.freepress.net.

Postman, Neil. (1985). *Amusing Ourselves to Death: Public Discourse in the Age of Show Business.* New York: Penguin Books.

Postman, Neil. (1999). *Building a Bridge to the 18th Century: How the Past Can Improve Our Future.* New York: Vintage Books.

Postman, Neil. (1993). *Teaching as a Subversive Activity.* New York: Vintage Books.

Postman, Neil (1993). *Technopoly: The Surrender of Culture to Technology.* New York: Vintage Books.

Project for Excellence in Journalism. Principles of Journalism. Available at http://www.journalism.org/resources/principles.

Rideau, Emile (1967). *The Thought of Teilhard de Chardin* (Rene Hague, Trans.). New York: Harper and Row.

Rosen, Jay (1996). *Getting the Connections Right: Public Journalism and the Troubles in the Press.* New York: The Twentieth Century Fund Press.

Rosen, Jay (1999). *What Are Journalists For?* New Haven, CT: Yale University Press.

Rosen, Jeffrey (2010, July 19). The web means the end of forgetting. *The New York Times.* Retrieved from http://www.nytimes.com/2010/07/25/magazine/25privacy-t2.ht… Downloaded August 2, 2010.

Roth, Robert J. (1962). *John Dewey and Self-Realization.* Englewood Cliffs, NJ: Prentice-Hall, Inc.

Rushkoff, Douglas (2005) Nieman Reports. (2005, Winter). Citizen Journalism and the BBC. *Neiman Reports: Citizen Journalism 59(4),* 12–15. Available at www.nieman.harvard.edu.

Sambrook, Richard (2005, Winter). Citizen Journalism and the BBC. *Nieman Reports: Citizen Journalism 59(4),* 12–15. Available at www.nieman.harvard.edu.

Samson, Paul R., and David Pitt. Eds. (1999). *The Biosphere and Noosphere Reader: Global Environment, Society and Change.* New York: Routledge.

Sanford, Bruce W. (1999). *Don't Shoot the Messenger: How Our Growing Hatred of the Media Threatens Free Speech for All of Us.* Lanham, MD: Rowman & Littlefield Publishers.

Schramm, Wilbur (1948). *Communications in Modern Society.* Urbana: University of Illinois Press.

Schramm, Wilbur (1973). *Men, Messages, and Media: A Look at Human Communication.* New York: Harper and Row.

Schudson, Michael (1973). *Discovering the News: A Social History of American Newspapers.* New York: Basic Books.

Schudson, Michael (1995). *The Power of News.* Cambridge, MA: Harvard University Press.

Schudson, Michael (2001, August). The objectivity norm in American journalism. *Journalism* 2(2), 149–170.

Schudson, Michael (2003). *The Sociology of News.* New York: W.W. Norton.

Serfain, Rafal (1991). "Noosphere, Gaia and the science of the biosphere." Environmental Ethics 10: 121-37 in *The Biosphere and Noosphere Reader,* Samson and Pitt, eds. (1999). London. Routledge.

Shirky, Clay (2009, March 13). Newspapers and thinking the unthinkable. [Web log post].

Sinclair, Upton (1920). *The Brass Check: A Study of American Journalism.* Pasadena, CA: Upton Sinclair.

Soleri, Paolo (1983). "Teilhard and the Esthetic" In *The Desire to Be Human: A Global Reconnaissance of Human Perspectives in the Age of Transformation.* Zonneveld, Leo, and Robert Muller, eds. (1983). Wassenaar, Netherlands: Mirananda Publishers.

Starr, Paul (2004). *The Creation of the Media: Political Origins of Modern Communications.* New York: Basic Books.

Starr, Paul (2009, March 4). Goodbye to the age of newspapers (hello to a new era of corruption). *The New Republic.* Retrieved from http://www.tnr.com/article/goodbye-the-age-newpapers-hello-... Downloaded February 25, 2010.

Stephens, Mitchell (1990, Spring). A short history of news: some surprising facts about the history of journalism. *Media and Values, 50.* Retrieved from http://www.medialit.org/reading_room/article409.html. Downloaded March 3, 2010.

Stephens, Mitchell (1997). *A History of News.* Fort Worth, TX: Harcourt Brace.

Stovall, James Glen (2004). *Web Journalism: Practice and Promise of a New Medium.* Boston: Pearson Education.

Strate, Lance (2004, Summer). 9: The New York school and communication studies; 12: Formal roots; 7: Media history; 13: Conclusions. *Communication Research Trends 23(2)* 19(6), 1–20. Downloaded September 12, 2008.

Strate, Lance (2008). Studying media as media: McLuhan and the media ecology approach. *Media Tropes 1,* 127–142.

Teilhard de Chardin, Pierre (1959). *The Phenomenon of Man.* London: Collins.

Teilhard de Chardin, Pierre (1962). *Human Energy.* New York: Harcourt Brace.

Teilhard de Chardin, Pierre (1966). *The Vision of the Past.* New York: Harper &Row.

Teilhard de Chardin, Pierre (1999). *The Human Phenomenon.* Brighton, UK: Sussex Academic Press.

Tracy, Christine M. (2005). "Examining an Emergent Medium: A Case Study of the Development of the TimesUnion.com." Doctoral dissertation, Rensselaer Polytechnic Institute, Troy, NY.

Tracy, Christine (2007). "The Trilogy: An Effective Rhetorical Strategy for Designing Digital Texts." *Language Arts Journal of Michigan,* Fall 2007, 36–41.

Tracy, Christine (2008). "Ecology and Democracy: Citizen Journalism in the Digital Age." *2008 Conference Proceedings of the Media Ecology Association.* Available at http://www.media-ecology.org/publications/MEA_proceedings/.

Tracy, Christine M. (2010). "Teilhard de Chardin and a technology of grace." *EME, the Journal of the Media Ecology Association, (8) 2.*

Tracy, Christine M. (2011). "A Quantum Exploration of the News Ecosystem." *EME, the Journal of the Media Ecology Association, (9) 3.*

Tracy, Christine M. and Robert K. Logan (2008). "A Biological Approach to the Rhetoric of Emergent Media." *2008 Conference Proceedings of the Media Ecology Association.* Available at http://www.media-ecology.org/publications/MEA_proceedings/.

Van Dusseldorp, Monique Roisin Scullion and Jan Bierhoff (1999, June). The future of the printed press: challenges in a digital world. *European Journalism Centre.* Retrieved from http://ejc.nl/hp/fpp/execsum.html. Downloaded April 30, 2005.

What is load shedding and why is it carried out? (n.d.) Retrieved January 24, 2011, from http://wiki.answers.com/!/What_is_load_shedding_and_why_is.... Downloaded on January 24, 2011.

Yates, Frances A. (1966). *The Art of Memory.* Chicago: The University of Chicago Press.

Zonneveld, Leo, and Robert Muller (1983). *The Desire to Be Human: A Global Reconnaissance of Human Perspectives in the Age of Transformation.* Wassenaar, Netherlands: Mirananda Publishers.

INDEX